Laboratory Manual on Concrete Technology

Laboratory Manual on Concrete Technology

Dr. Hemant Sood
Assistant Professor, Department of Civil Engineering,
Technical Teachers' Training Institute, Chandigarh

Dr. L.N. Mittal
Ex-Professor & Head, Curriculum Development Centre,
Technical Teachers' Training Institute, Chandigarh

Prof. P.D. Kulkarni
Ex-Professor & Principal,
Technical Teachers' Training Institute, Chandigarh

CBSPD

CBS Publishers & Distributors Pvt Ltd

New Delhi • Bengaluru • Chennai • Kochi • Kolkata • Lucknow • Mumbai
Hyderabad • Jharkhand • Nagpur • Patna • Pune • Uttarakhand

Laboratory Manual on Concrete Technology

ISBN: 978-81-239-0941-7

First Edition: 2003
Reprint: 2008, 2009, 2010, 2011, 2013, 2014, 2016, **2025**

Published by **Satish Kumar Jain** and produced by **Varun Jain** for

CBS Publishers & Distributors Pvt Ltd
4819/XI Prahlad Street, 24 Ansari Road, Daryaganj, New Delhi 110 002, India.
Ph: 011-23266838, 23289259 Website: www.cbspd.com
e-mail: delhi@cbspd.com
Corporate Office: 204 FIE, Industrial Area, Patparganj, Delhi 110 092
Ph: 011-4934 4934 Fax: 011-4934 4935
e-mail: publishing@cbspd.com; publicity@cbspd.com

Branches

- **Bengaluru:** Seema House 2975, 17th Cross, KR Road, Banasankari 2nd Stage, Bengaluru 560 070, Karnataka, India
  Ph: +91-80-26771678/79 Fax: +91-80-26771680 e-mail: bangalore@cbspd.com
- **Chennai:** 18/8B, Subbarayan Street, Shenoy Nagar, Chennai 600 030, Tamil Nadu, India
  Ph: +91-44-42032115, 26681266 e-mail: chennai@cbspd.com
- **Kochi:** 42/1325, 1326, Power House Road, Opp KSEB, Power House, Ernakulum Kochi 682 018, Kerala, India
  Ph: +91-484-4059061-65,67 Fax: +91-484-4059065 e-mail: kochi@cbspd.com
- **Kolkata:** 147, Hind Ceramics Compound, 1st Floor, Nilgunj Road, Belghoria, Kolkata-700056, West Bengal, India
  Ph: +033-25633055, 033-25633056 e-mail: kolkata@cbspd.com
- **Lucknow:** Basement, Khushnuma Complex, 7 Meerabai Marg (Behind Jawahar Bhawan), Lucknow-226001, UP, India
  Ph: +0522-4000032 e-mail: tiwari.lucknow@cbspd.com
- **Mumbai:** PWD Shed, Gala no 25/26, Ramchandra Bhatt Marg, Next to JJ Hospital Gate no. 2, Opp. Union Bank of India, Noorbaug, Mumbai-400009, Maharashtra, India
  Ph: 022-66661880/89 e-mail: mumbai@cbspd.com

Representatives

- Hyderabad 0-9885175004
- Jharkhand 0-9811541605
- Nagpur 0-8692091830
- Patna 0-9334159340
- Pune 0-9664372571
- Uttarakhand 0-9716462459

Printed at Chaman Enterprises, Daryaganj, New Delhi, India

Foreword

The conduct of practical work in the laboratories of the polytechnic in a scientific manner is very important if the polytechnic education is really to be job-oriented. At present the practicals in the laboratory are usually conducted in a routine manner, seldom keeping the objectives of the course in mind. Students are asked to perform experiments but are not asked to evaluate the results. There are no useful manuals available to them from where they can study all the relevant material for conducting practical experimentation in a systematic way. This institute has been engaged in preparing suitable manuals with the intention of stream-lining the laboratory procedures and as a result it has brought out this publication.

The material prepared has been thoroughly discussed by the experienced teachers, field engineers and by the experts of the Bureau of Indian Standards before publication. The manual has been thoroughly revised and updated in time with the latest BIS Specifications. New experiments on workability, Blain's Air Permeability tested, accelerated curing test etc. and a new chapter of latest Non-Destructive Techniques have also been included in this manual which makes it complete document for laboratory testing in Concrete Technology.

It is hoped that this manual will be very useful to polytechnic teachers, students and working professionals in carrying out practical work in concrete laboratory as per BIS Standards.

Dr. S.K. Bhattacharya
Principal

Preface

Concrete is one of the important materials of construction in Civil Engineering. The durability of the structure depends upon the care with which ingredients of concrete are selected mixed, laid, vibrated and cured. This laboratory manual deals mainly with the selection of the ingredients and preparation of ordinary concrete mix as per latest BIS code.

The Graduate Civil Engineer must possess the following knowledge and abilities.

(i) He must know how to store and stock the materials required for concrete.

(ii) He must know the methods of sampling, so as to have representative sample of cement, aggregate and concrete from bulk. He must know how much quantity is required for each test.

(iii) He must be able to perform the tests himself according to the standard procedure laid down.

(iv) He must be able to evaluate the results of the experiments and draw his own conclusions about the suitability for use in concrete. For this he must be fully aware of the requirements of good cement, aggregate and concrete.

(v) He must have the knowledge and ability to perform non-destructive tests on hardened concrete.

The information necessary to enable the engineers to discharge his duties efficiently, has been designed, collected and presented in this manual in suitable parts. The necessary theory and instructions for use in the laboratory have also been given at the beginning. With this manual in hand, any engineer can study, and perform the experiments and evaluate the results.

Indian standard codes have been very frequently referred to for the preparation of this manual, for which the authors gratefully acknowledge the contribution of BIS, New Delhi.

Hemant Sood
L.N. Mittal
P.D. Kulkarni

About the Book & Authors

Concrete is one of the most important materials of construction. Durability of structures depends upon the quality of concrete which includes selection of appropriate aggregates, their batching, proportioning, mixing, transportation, placement and curing. Therefore, it is of paramount importance for engineers and technicians to understand the properties and procedure of testing such properties which govern the strength and quality of concrete. This laboratory manual deals mainly with the testing of various ingredients and concrete as per BIS specifications. The manual, besides testing aggregates, cement and concrete, also deals with various non-destructive testing techniques. The manual has been written strictly in accordance with BIS codes and is expected to be very useful to practising engineers besides technical teachers and students undergoing educational training programmes in various institutions and polytechnics.

Dr. Hemant Sood is Assistant Professor in the Department of Civil Engineering at the Technical Teachers' Training Institute, Chandigarh where he is involved with the development of instructional material, both print and non-print, staff development, curriculum development and R&D projects.

Hemant Sood was actively involved in the development of Continuing Education Package on "Concrete Mix Design" published by Indian Society of Technical Education. He has undertaken number of consultancy projects in the field of concrete technology both for Government as well as private sector. He has also developed series of video films on testing of cement, earthmoving equipment and construction of flexible pavement. He is B.E. (Civil) with Honours and M.E. in Highway Engineering from Punjab Engineering College, Chandigarh. He did his Ph.D. from Kurukshetra University. Before joining TTTI, Chandigarh in December 1988, he served Rail India Technical and Economic Services Limited (RITES), Government of India Enterprise under Ministry of Railways as Assistant Manager for a period of one and a half years where he was involved in the designing of number of minor highway bridges besides strengthening the existing Agra-Jaipur National Highway. He has contributed number of papers in national journals and participated in International Conference on "High Performance Concrete" at Shah Alam, Malaysia in 1995. He is also a member of many professional bodies in India. He has also written a Laboratory Manual on 'Testing of Engineering Materials' published by New Age International (P) Limited, New Delhi, which was reprinted in 2001.

Dr. L.N. Mittal worked as Professor & Head, Curriculum Development Centre at TTTI, Chandigarh. He joined teaching profession in 1962. He has long experience of teaching at the polytechnic and at TTTI, Chandigarh. Dr. Mittal went to Netherlands and UK on Fellowship Training in 1974-75. He participated in World Conference on Cooperative Education at Edinburgh in 1985 and Kingston (Jamaica) in 1995 and was honoured to chair technical sessions at both these conferences. He has contributed over 30 papers in national and international journals. Dr. Mittal is a member of many professional bodies in India and abroad and is also an ISTE National Award Winner for presenting a paper on the "Concept of Cooperative Education". He has number of Academic Awards to his credit during his studenthood and later in his professional career.

Prof. P.D. Kulkarni graduated from the College of Engineering, Pune in 1953 in Civil Engineering. He completed his two years' professional training for technical teachers in Germany in 1963-66.

Having served in the Polytechnic Education System for 12 years from 1958-1969, he joined the Technical Teachers' Training Institute, Chandigarh since 1969 and retired in 1992. He served the institute both as a Professor and Principal.

He was UNESCO expert in Bangladesh for nine months for the development of Polytechnic Education. He was appointed as head of the National Project Implementation Unit of the Government of India for launching the World Bank assisted project for strengthening polytechnic education in India.

At present he is educational consultant to the State Board of Technical Education, Maharashtra, Mumbai, Indo-German Tool Room, Aurangabad and the Institute of Management Development & Research, Pune.

He was the member of the expert committees appointed by the Govt. of India for Vocationalization of Secondary Education; Direct Central Assistance; revision of pay-scales of technical teachers.

He has contributed to the promotion of knowledge of Theories of Learning, Instructional Design, Curriculum Development, Institutional Planning and State Level Educational Planning, System Thinking and Independent Study Techniques among teachers and managers of technical education in the country.

Contents

CHAPTER 1

Specifications for Cement

1.1 INTRODUCTION TO CEMENT

Cement is by far the most important constituent of concrete, in that it forms the binding medium for the discrete ingredients. Made out of naturally occurring raw materials and sometimes blended or interground with industrial wastes, cements come in various types and chemical compositions. For general concrete constructions, the IS:456 permits the use of the following types of cement, subject to the approval of the engineer-in-charge:

(a) Ordinary or low heat Portland cement conforming to IS:269.

(b) Rapid hardening Portland cement conforming to IS:8041.

(c) Portland slag cement conforming to IS:455.

(d) Portland pozzolana cement conforming to IS:1489.

(e) High strength Ordinary Portland cement conforming to IS:8112.

(f) Super-sulphated cement conforming to IS:6909.

(g) High strength Ordinary Portland cement conforming to IS:12269.

In addition to these, high alumina cement conforming to IS:6452 and hydrophobic cement conforming to IS:6909 can be used under special circumstances with necessary precautions. In case of Portland slag cement, the slag content should not be more than 50 per cent.

The detailed specifications of the most commonly used cements are given in Table 1.1.

Specifications for cement

Portland cements are manufactured in controlled conditions. For engineers, physical properties of cement are relevant as acceptance criteria. These are, therefore summarised in Table 1.1.

Table 1.1. Specifications for physical properties of Portland cements

S. No. Properties	33 grade Ordinary Portland cement (IS:269)	Rapid hardening Portland cement (IS:8041)	Low heat cement (IS:269)	Super sulphated cement (IS:6909)	Portland pozzolana cement (IS:1489)	43 grade Ordinary Portland cement (IS:8112)	53 grade Ordinary Portland cement (IS:12269)
1. Fineness							
(a) Residue by weight on IS Sieve 90 micron not to exceed per cent	10	—	—	—	—	—	—
(b) Specific surface (cm^2/g) by air permeability method not less than	2250	3250	3200	4000	3000	2250	2250
2. Setting time (in min.)							
(a) Initial setting time not less than	30	30	60	30	30	30	30
(b) Final setting time not more than	600	600	600	600	600	600	600
3. Compressive strength (kg/cm^2) of 1 : 3 cement mortar cube :							
• at 1 day (24 h + 30 min) not less than	—	160	—	—	—	—	—
• at 3 days (72 h + 1 h) not less than	160	275	100	150	—	230	270
• at 7 days (168 h + 2 h) not less than	220	—	160	220	220	330	370
• at 28 days (672 h + 4 h) not less than	—	—	350	300	310	430	530
Note : For compressive strength test, standard sand specified in IS:650 (revised) should be used.							
4. Soundness							
By Le Chatelier method specimen shall not have an expansion of more than (mm)	10	10	10	5 (cold expansion)	10	10	10 (unaerated) 5 (aerated)
By Autoclave method specimen shall not have expansion of more than (per cent)	0.8	0.8	0.8	—	0.8 (aerated)	0.8	0.8 (unaerated) 0.6 (aerated)
Note : As regards soundness test, in the event of cement failing to comply with the above requirements, a further test shall be made by Le-Chatelier Method, from another portion of the same sample, after aeration, by spreading out to a depth of 75 mm at a relative humidity of 50 to 80 per cent for a total period of 7 days, when the expansion of cements mentioned above shall not be more than 5 mm.							
5. Maximum heat of hydration							
• At 7 days not more than	—	—	65.0	—	—	—	—
• At 28 days not more than	—	—	75.0	—	—	—	—
6. Drying shrinkage (max. percentage)	—	—	—	—	0.15	—	—

Table 1.2. Chemical requirements for different types of cements

S. No.	Characteristics	33 grade ordinary Portland cement (IS:269)	43 grade ordinary Portland cement (IS:8112)	53 grade ordinary Portland cement (IS:12269)	Rapid hardening Portland cement (IS:8041)	Low heat Portland cement (IS:269)	Portland pozzolana cement (IS:1489)	Super sulphated cement (IS:6909)
1.	Insoluble residue per cent by mass (not more than)	2	2	2	2	2	$^{*}x + \frac{2.0(100 - x)}{100}$	4
2.	Magnesia, per cent by mass (not more tan)	6	6	6	6	6	6	10
3.	Total loss on ignition (not more than)	5	4	4	5	5	5	—

* where 'x' is the declared percentage of pozzolana in the given Portland Pozzolana cement.

CHAPTER 2

Specifications and Data for Coarse & Fine Aggregates and Water

Aggregate constitutes about 75 per cent of the bulk of concrete. The properties of aggregate, therefore, considerably affect the quality of concrete. Among many properties, shape, size, presence of impurities, crushing strength and grading of aggregate are very important and following paragraphs lay down the standard specifications for good quality aggregate.

For ordinary concrete, certain data such as fineness-modulus, specific gravity, bulk density, surface water in aggregate, water absorption in aggregate, bulking of sand are required to be assumed in design. Practical data based on experience has been collected and are reproduced in the ensuing paragraphs.

2.1 INSPECTION OF AGGREGATE (IS:383)

A. Trade groups of rocks used as concrete aggregate

Table 2.1

Name of rock	*Trade group*
Igneous rock	Granite group, gabbro group, rhyolite group, basalt group
Sedimentary rock	Sandstone group, limestone group
Metamorphic rock	Granulite and gneiss group, schist group, marble group
Derivative rock	Residual group

Correct identification of a rock and its placing under the appropriate trade group should be left to the Geological Survey of India or any competent geo-logist.

B. Particle shape

To avoid lengthy description of the external characteristics of mineral aggregate, a similar system as given in Tables 2.2 and 2.3 has been devised to define both particle shape and surface characteristics.

Table 2.2. Particle shape

S.	*Classification*	*Description*	*Illustrations of characterstic specimens*	*Examples*
1.	Rounded	Fully water worn or completely shaped by attrition	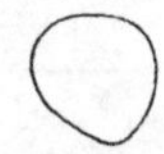	River or sea shore gravels, desert, sea-shore and wind-blown sands
2.	Irregular or partly rounded	Naturally irregular or partly shaped by attrition, and having rounded edges		Pit-sands and gravels; land or dug-flints; cuboid rock
3.	Angular	Possessing well-defined edges formed at the inter-section of roughly planar faces	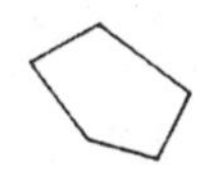	Crushed rocks of all types; talus; screes
4.	Flaky	Material, usually angular of which the thickness is small relative to the width and/or length		Laminated rocks

C. Surface characteristics of aggregates

Table 2.3. Surface characteristics of aggregates

Group	*Surface texture*	*Example*
1.	Glassy	Black flint
2.	Smooth	Chert, slate, marble, some rhyolite
3.	Granular	Sand stone; Oolites
4.	Crystalline	*Fine* : Basalt; trachyte; keratophyre *Medium* : Dolerite; granophyre; granulite; microgra-nite; some limestones; many dolomites *Coarse* : Gabbro; gneiss; granite; granodiorite; syenite
5.	Honeycombed and porous	Scoriac, pumice, trass

2.2 LIMITS OF DELETERIOUS MATERIALS (IS:383)

The maximum quantity of deleterious material shall not exceed the limits specified in Table 2.4 when tested in accordance with IS:2386. However, the engineer-in-charge at his discretion may relax some of the limits as a result of some further tests and evidence of some satisfactory performance tests.

2.3 PERCENTAGE OF SILT

Percentage of silt by volume after three hours of setting should not exceed 6 per cent (IS has not given any specification for this).

Table 2.4. Limits of deleterious materials (IS:383)

S. No.	Deleterious substance	Method of test	Fine aggregate percentage by weight, max.		Coarse aggregate percentage by-weight, max.	
			Uncrushed	Crushed	Uncrushed	Crushed
1	2	3	4	5	6	7
(i)	Coal and lignite	IS:2386 (Part-II)	1.00	1.00	1.00	1.00
(ii)	Clay lumps	IS:2386 (Part-II)	1.00	1.00	1.00	1.00
(iii)	Materials finer than 75 micron IS sieve	IS:2386 (Part-I)	3.00	15.00	3.00	3.00
(iv)	Soft fragments	IS:2386 (Part-II)	—	—	3.00	—
(v)	Shale	IS:2386 (Part-II)	1.00	—	—	—
(vi)	Total of percentages of all deleterious materials (except mica) including S.No. (i) to (v) for column 4, 6 and 7 and S.No. (i) and (ii) for Col. 5 only	—	5.00	2.00	5.00	5.00

Note: 1. The sum of the percentages of all deleterious materials shall not exceed five.
2. The presence of mica in the fine aggregates has been found to reduce considerably the durability and compressive strength of concrete and further investigations are underway to determine the extent of the deleterious effect of mica. It is advisable, therefore, to investigate mica content of fine aggregate and make suitable allowances for the possible reduction in the strength of concrete or mortar.
3. Aggregate shall not contain harmful organic impurities (tested in accordance with IS:2386 (Part-II) in sufficient quantities to affect adversely the strength or durability of concrete. A fine aggregate which fails in the test for organic impurities may be used, provided that, when tested for the effect of organic impurities on the strength of mortar, the relative strength at 7 and 28 days, reported in accordance with 7 of IS:2386 (Part-VI) is not less than 95 per cent.

2.4 AGGREGATE CRUSHING VALUE (IS:383)

The aggregate crushing value gives a relative measure of the resistance of an aggregate to crushing under a gradually applied compressive load with aggregate of 'aggregate crushing value' 30 or higher, the result may be anomalous, and in such cases the 'ten per cent fines value' should be determined instead. The aggregate crushing value, when determined in accordance with IS:2386 (Part-IV) shall not exceed 45 per cent for aggregate used for concrete other than for wearing surfaces, and 30 per cent for concrete for wearing surfaces, such as runways, roads and pavements.

2.5 AGGREGATE IMPACT VALUE (IS:383)

As an alternative to aggregate crushing value test, the aggregate impact value may be determined in accordance with the method specified in IS:2386 (Part IV). The aggregate impact value shall not exceed 45 per cent by weight for aggregates used for concrete other than for wearing surfaces and 30 per cent by weight for concrete for wearing surfaces, such as runways, roads and pavements.

2.6 AGGREGATE ABRASION VALUE (IS:383)

The abrasion value of aggregates, when tested in accordance with the method specified in IS:2386 (Part IV) using Los Angeles Machine, shall not exceed the following values :

(a) For aggregates to be used in concrete for wearing surfaces — 30 per cent
(b) For aggregates to be used in other concrete — 50 per cent

2.7. SOUNDNESS OF AGGREGATE (IS:383)

For concrete liable to be exposed to the action of frost, coarse and fine aggregates shall pass a sodium or magnesium sulphate accelerated soundness test specified in IS:2386 (Part V), the limits being set by agreement between the purchaser and the supplier, except that aggregates failing in the accelerated soundness test may be used if they pass a specified freezing and thawing test satisfactory to the user.

Note : As a general guide, it may be taken that the average loss of weight after 5 cycles shall not exceed the following:

(a) *For fine aggregate* : 10 per cent when tested with sodium sulphate (Na_2SO_4) and 15 per cent when tested with magnesium sulphate ($MgSO_4$).
(b) *For coarse aggregate* : 12 per cent when tested with sodium sulphate (Na_2SO_4) and 18 per cent when tested with magnesium sulphate ($MgSO_4$).

2.8 SIZE AND GRADING OF AGGREGATES (IS:383)

A. Fine aggregate

The grading of fine-aggregates, when determined as described in IS:2386 (Part I) shall be within the limits given in the Table 2.5 and shall be described as fine aggregates Grading Zones I, II, III and IV. Where the grading falls outside the limits of any particular grading zone of sieves other than 600 micron IS seive by a total amount not exceeding 5 per cent, it shall be regarded as falling within that grading zone. This tolerance shall not be applied to percentage passing 600-micron IS Sieve or to percentage passing any other sieve size on the coarser limit of grading zone I or finer limit of grading zone IV.

Table 2.5. Fine aggregates (see also Fig. 2.1 and 2.2)

IS sieve	*Percentage passing by weight*			
Designation	*Grading zone I*	*Grading zone II*	*Grading zone III*	*Grading zone IV*
10 mm	100	100	100	100
4.74 mm	90-100	90-100	90-100	95-100
2.36 mm	60-95	75-100	85-100	95-100
1.18 mm	30-70	55-90	75-100	90-100
600 micron	15-34	35-59	60-79	80-100
300 micron	5-20	8-30	12-40	15-50
150 micron	0-10	0-10	0-10	0-15

Note: 1. For crushed stone sands, permissible limit on 150 micron IS sieve is increased to 20 per cent. This does not effect the 5 per cent limit permitted as prescribed above applying to other sieve sizes.

2. Fine aggregate complying with the requirements of any grading zone in this table is suitable for concrete but the quality of concrete produced will depend upon a number of factors including proportions.

3. Where concrete of high strength and good durability is required, fine aggregates conforming to any one of the four grading zones may be used, but the concrete mix should be properly designed. As the fine aggregate grading becomes progressively finer, i.e. from grading zone I to IV, the ratio of fine aggregate to

coarse aggregate should be progressively reduced. The most suitable fine to coarse ratio to be used for any particular mix will, however, depend upon the actual grading, particle shape and surface texture of both fine and coarse aggregates.

4. It is recommended that fine aggregates conforming to grading zone IV should not be used in reinforced concrete unless tests have been made to ascertain the suitability of proposed mix proportions.

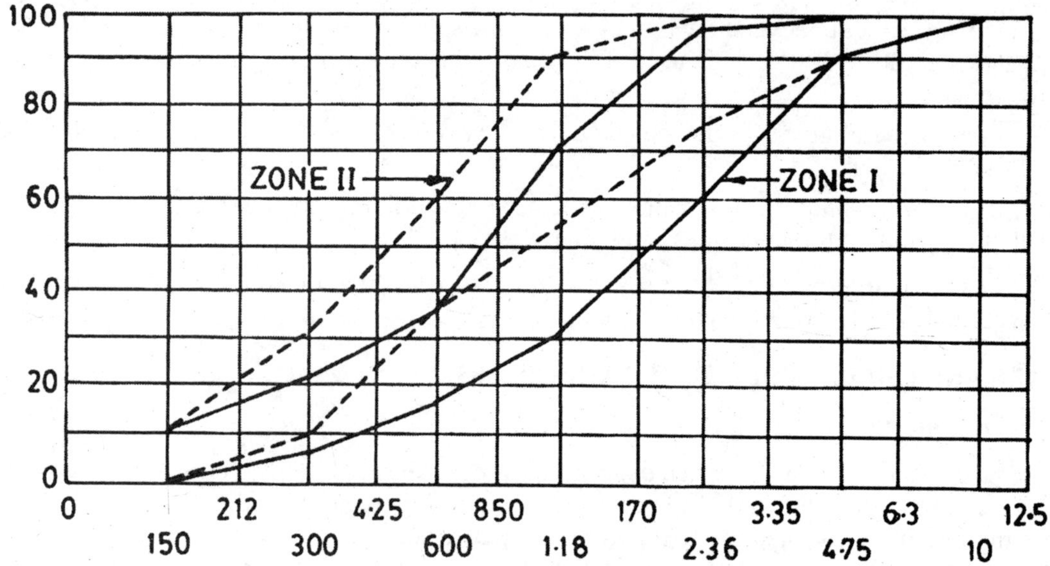

Fig. 2.1. Grading limits for zone I and zone II of fine aggregate.

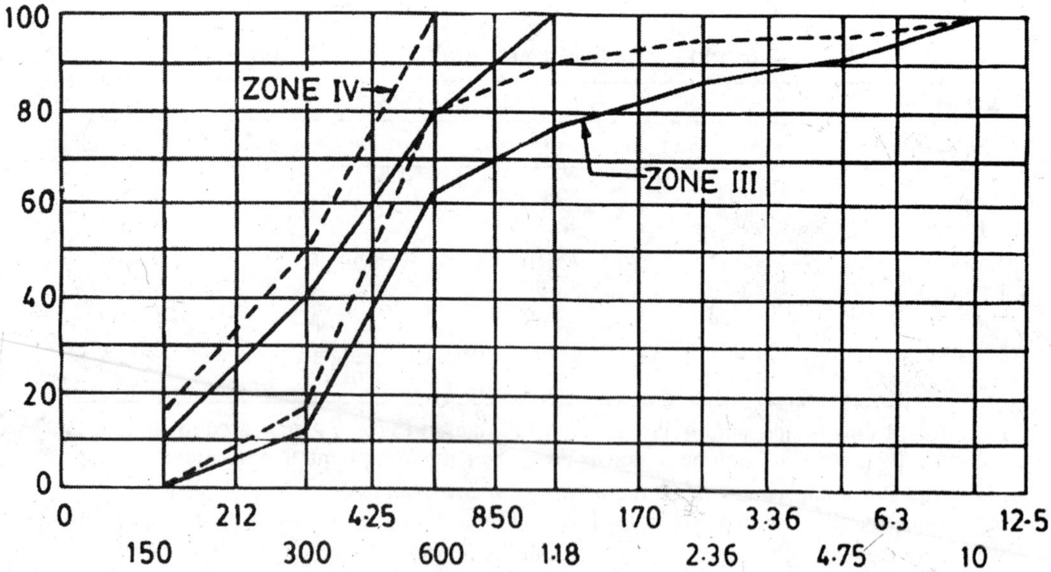

Fig. 2.2. Grading limits for zone III and zone IV of fine aggregate.

B. Single size and graded coarse aggregate

The coarse aggregate both graded and single sized shall be supplied in nominal sizes given in Table 2.6. For any one of the nominal sizes, the proportions of other sizes, as determined by the method described in IS:2386 (Part I) shall also be in accordance with the Table 2.6.

Table 2.6. Single size and graded coarse aggregate (see also Fig 2.3a and Fig. 2.3b).

IS sieve Designation	Percentage passing for single size aggregate of nominal size						Percentage passing for graded aggregate of nominal size			
	63 mm	40 mm	20 mm	16 mm	12.5 mm	10 mm	40 mm	20 mm	16 mm	12.5 mm
80 mm	100	—	—	—	—	—	100	—	—	—
63 mm	85-100	100	—	—	—	—	—	—	—	—
40 mm	0-30	85-100	100	—	—	—	95-100	100	—	—
20 mm	0-5	0-20	85-100	100	—	—	30-70	95-100	100	100
16 mm	—	—	—	85-100	100	—	—	—	90-100	—
12.5 mm	—	—	—	—	85-100	100	—	—	—	90-100
10 mm	0-5	0-5	0-20	0-30	0-45	85-100	10-35	25-55	30-70	40-85
4.75 mm	—	—	0.5	0.5	0-10	0-20	0-5	0-10	0-10	0-10
2.36 mm	—	—	—	—	—	0.5	—	—	—	—

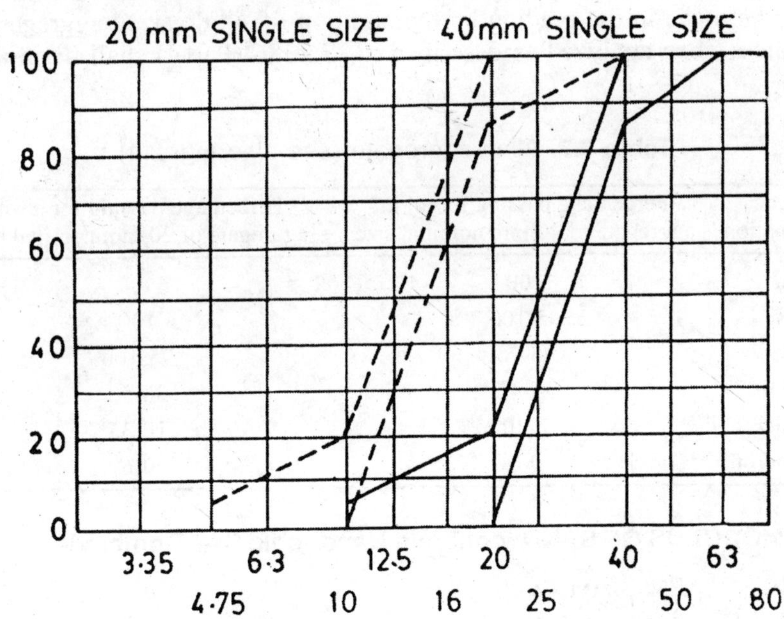

Fig. 2.3(a). Grading limits for coarse aggregate 20 mm and 40 mm single size.

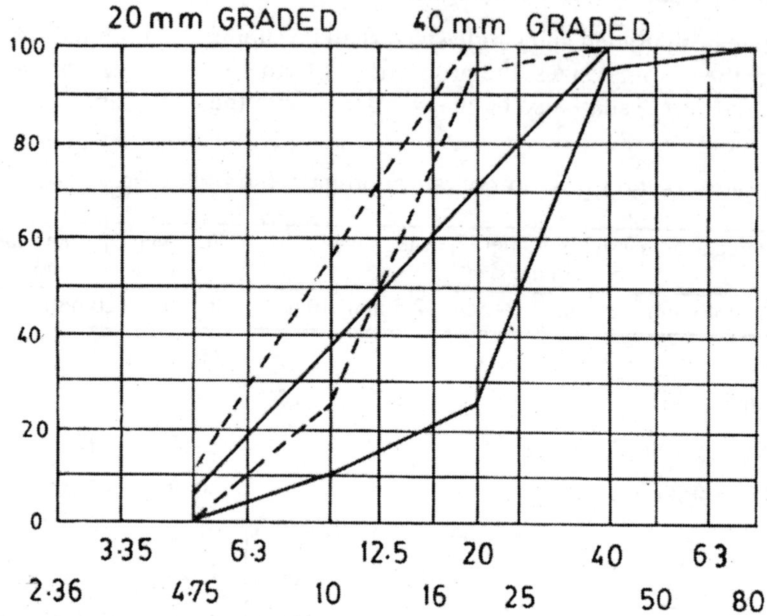

Fig. 2.3(b). Grading limits for coarse aggregate 20 mm and 40 mm graded.

C. All-in Aggregate

If combined aggregates are available they need not be separated into fine and coarse, but necessary adjustments may be made in the grading by the addition of single sized aggregates. The grading of all in-aggregates, when analyzed as described in IS:2386 (Part I) shall be in accordance with Table 2.7.

Table 2.7. All-in-aggregate (see also Fig. 2.4)

IS sieve Designation	Percentage passing for all in aggregate of 40 mm nominal size	Percentage passing for all-in aggregate of 20 mm nominal size
80 mm	100	—
40 mm	95-100	100
20 mm	45-75	95-100
4.75 mm	25-45	30-50
600 micron	8-30	10-35
150 micron	0-6	0-6

2.9 FINENESS MODULUS OF SAND (concrete Handbook, CAI, Bombay)

- Fine sand FM 2.20 to 2.60
- Medium sand FM 2.60 to 2.90
- Coarse sand FM 2.90 to 3.20

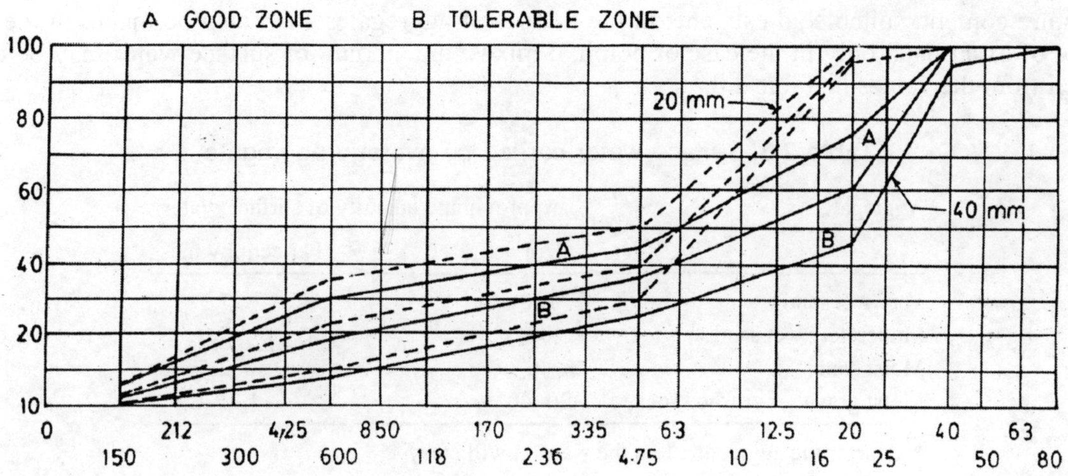

Fig. 2.4. Grading limits for all-in-aggregate.

2.10 SPECIFIC GRAVITY (Concrete Handbook, CAI, Bombay)

• Cement	3.15
• Medium sand	2.00
• Granite	2.80
• Gravel	2.66
• Sand	2.65

2.11 BULK DENSITY (kg/litre) (Concrete Handbook, CAI, Bombay)

• Cement	1.44
• River Sand :	
Fine	1.44
Medium	1.52
Coarse	1.60
Beach or river shingle	1.60
Broken stone	1.60
Stone screening	1.44
Broken granite	1.68

2.12 SURFACE WATER CARRIED BY AGGREGATE (IS:456)

In order to maintain the constant water cement ratio it is necessary to determine the moisture contents in both fine and coarse aggregates as frequently as possible. The frequency to check the moisture contents of aggregates is determined by the engineer-in-charge according to weather conditions. The amount of the added water shall be adjusted to compensate for any observed variations in the moisture contents. For the determination of moisture content in the aggregates, IS:2386 (Part III) may be referred to. To allow for the variation in mass of aggregate due to variation in their moisture content, suitable adjustments in the masses of aggregates shall also be made. In the absence of exact data, only in the case of nominal mixes, the amount of surface water may be estimated from the values in Table 2.8.

Table 2.8. Surface water carried by average aggregate

Aggregate	Approximate quantity of surface water	
	l/m^3	Per cent by mass
Very wet sand	120	7.5
Moderately wet sand	80	5.0
Moist sand	40	2.5
Moist gravel or crushed rock*	20–40	1.25–2.5

* Coarser the aggregate, less the water it will carry.

2.13 ABSORPTION OF THE WATER BY DRY AGGREGATE (IS:456)

Table 2.9. Absorption of the water by dry aggregate

Aggregate	*Moisture absorbed by wt. of aggregate*
Average sand	1.0 per cent
Pebbles and crushed limestone	1.0 per cent
Trap-rock and granite	0.5 per cent
Porous sandstone	7.0 per cent
Very light and porous aggregate	25 per cent

2.14. BULKING OF SAND WITH VARIOUS MOISTURE CONTENTS (Concrete Handbook, CAI Bombay)

Table 2.10. Bulking of sand

Moisture	Percentage Bulking in		
per cent	Fine sand	Medium sand	Coarse sand
1	16	8	6
2	26	16	12
3	32	22	15
4	36	27	17
5	30	29	18
6	37	28	18
8	35	26	16
10	32	22	12
12	28	19	8
15	22	12	2
17	10	7	0
20	0	0	0
27	0	0	0

2.15 FLAKINESS INDEX AND ELONGATION INDEX

Indian standards has not laid down limits for these indices. But British standard BS 63 lays down maximum limit of 40 per cent for aggregate size between 2" (50 mm) to 1" (25 mm) and 35 per cent for aggregate sizes between 1" (25 mm) to ¼" (6.3 mm). Minimum number of pieces to be tested when determining the index as per B.S. 63 are as under :

Nominal size	No. of pieces
1¼" (31.5 mm) to 2" (50 mm)	300
3/4" (20 mm) to 1" (25 mm)	290
3/4" (20 mm) and under	100

BS 1241 stipulates a flakiness index of not greater than 30 per cent.

2.16 SPECIFICATIONS FOR WATER (IS:456)

(i) *Water* : Water used for mixing and curing shall be clean and free from injurious amounts of oil, acids, alkalis, salts, sugar, organic materials, or other substances that may be deleterious to concrete or steel. Potable water is generally considered satisfactory for mixing concrete. As a guide, the following concentrations represent the maximum permissible values :

(a) To neutralize 100 ml sample of water, using phenolphthalein as an indicator, it should not require more than 5 ml of 0.02 normal NaOH. The details of test are given in 8.1 of IS:3025 (Past 22).

(b) To neutralize 100 ml sample of water, using mixed indicator, it should not require more than 25 ml of 0.02 normal H_2SO_4. The details of test shall be as given in 8 of IS:3025 (Past 23).

(c) Permissible limits for solids when tested in accordance with IS : 3025 shall be as given in Table 2.11.

Table 2.11. Permissible limit for solids

S.No.	*Teste as per*	*Permissible limit, max*
(i) Organic	IS 3025 (Part 18)	200 mg/l
(ii) Inorganic	IS 3025 (Part 13)	3000 mg/l
(iii) Sulphates (as SO_4)	IS 3025 (Part 24)	400 mg/l
(iv) Chlorides (as Cl)	IS 3025 (Part 32)	2000 mg/l for concrete not containing embedded steel and 500 mg/l for reinforced concrete work
(v) Suspended matter	IS 3025 (Part 17)	2000 mg/l

(ii) In case of doubt regarding development of strength, the suitability of water for making concrete shall be ascertained by the compressive strength and initial setting time tests specified in (iv) and (v) below.

(iii) The sample of water taken for testing shall represent the water proposed to be used for concreting, due account being paid to seasonal variation. The sample shall not receive any treatment before testing other than that envisaged in the regular supply of water proposed for use in concrete. The sample shall be stored in a clear container previously rinsed out with similar water.

(iv) Average 28 days compressive strength of at least three 15 cm concrete cubes prepared with water proposed to be used shall not be less than 90 per cent of the average of strength of three similar concrete cubes prepared with distilled water. The cubes shall be prepared, cured and tested in accordance with the requirements of IS:516.

(v) The initial setting time of test block made with the appropriate cement and the water proposed to be used shall not be less than 30 minutes and shall not differ by $\pm$ 30 minutes from the initial setting time of control test block prepared with the same cement and distilled water. The test blacks shall be prepared and tested in accordance with the requirements of IS : 4031 (Part 5).

(vi) The pH value of water shall be not less than 6.

Sea water

Mixing or curing of concrete with sea water is not recommended because of presence of harmful salts in sea water. Under unavoidable circumstances, sea water may be used for mixing or curing in plain concrete with no embedded steel after having given due consideration to possible disadvantages and precautions including use of appropriate cement system.

Water found satisfactory for mixing is also suitable for curing concrete. However, water used for curing should not produce any objectionable stain or unsighty deposit on the concrete surface. The presence of tannic acid or iron compounds is objectionable.

CHAPTER 3

Specifications and Data for Concrete (Adopted from IS:456–2000)

3.1 GRADES OF CONCRETE

The concrete shall be in fifteen grades designated as :

M_{10}, M_{15}, M_{20}, M_{25}, M_{30}, M_{35}, M_{40}, M_{45}, M_{50}, M_{55}, M_{60}, M_{65}, M_{70}, M_{75}, M_{80} as per IS:456.

Note : (i) In the designation of a concrete mix, letter M refers to the mix and the number to the specified characteristic compressive strength of 150 mm cube at 28 days, expressed in N/mm^2.

(ii) For concrete of compressive strength greater than M_{55}, design parameters given in the standard may not be applicable and the values may be obtained from specialized literatures and experimental results.

3.2 STRENGTH REQUIREMENTS OF CONCRETE

Where ordinary Portland cement conforming to IS:269 or Portland blast-furnace slag cement conforming to IS:455 is used, the compressive strength requirements for various grades of concrete shall be as given in Table 3.1. Where rapid hardening Portland cement is used, the 28 day compressive strength requirements specified in Table 3.1 shall be met at 7 days. Where other cements are usedc, the engineer-in-charge shall specify the corresponding requirements preferably on the basis of preliminary tests.

The strength requirements specified in Table 3.1 shall apply to both controlled concrete and ordinary concrete. Preliminary test need not however be made in the case of ordinary concrete.

Where the strength of a concrete mix, as indicated by tests, lies in between the strength for any grades, specified in Table 3.1, such concrete shall be classified for all purpose as a concrete belonging to the lower of the grades between which its strength lies.

Table 3.1. Strength requirement of concrete [all values in kg/cm² (N/mm²)]

Group	*Grade designation*	*Specified characteristic compressive strength of 150 mm cube at 28 days in N/mm²*
Ordinary Concrete	M 10	10
	M 15	15
	M 20	20
Standard Concrete	M 25	25
	M 30	30
	M 35	35
	M 40	40
	M 45	45
	M 50	50
	M 55	55
High Strength Concrete	M 60	60
	M 65	65
	M 70	70
	M 75	75
	M 80	80

Note :

1. In the designation of concrete mix M refers to the mix and the number to the specified compressive strength of 150 mm size cube at 28 days, expressed in N/mm^2.
2. For concrete of compressive strength greater than M 55, design parameters given in the standard may not be applicable and the values may be obtained from specialized literatures and experimental results.

1. (a) At the site, with the approval of the engineer-in-charge, 10 cm cubes may be used in place of 15 cm cubes provided the maximum nominal size of aggregate does not exceed 20 mm. Even the use of 15 cm cubes should normally be restricted to concrete having a maximum nominal size of aggregate not exceeding 40 mm. Where concrete with aggregates larger than 40 mm size is required to be tested, the size of cubes is specified by the engineer-in-charge, keeping in view that generally, the length of side of the cubes should be about four times the maximum nominal size of aggregate in the concrete constituting the cube-specimen.

   (b) *Strength in relation to size of the cube* : Where 10 cm cubes are used, the values obtained from tests on 10 cm cubes shall be reduced by 5 per cent of the value determined from the tests, in order to give equivalent strength for 15 cm cubes. When cubes larger than 15 cm are adopted then the values of strength obtained are increased by 10 per cent in order to give equivalent strength for 15 cm cubes.

   (c) *Cylinder strength* : Compressive strength tests may with the approval of engineer-in-charge be conducted on 15 cm diameter and 30 cm high cylinders in accordance with IS:516 instead of on cubes. Where cylinder strength figures are adopted the compressive strength figures given above shall be modified according to the formula:

   Minimum cylinder compressive strength required = 0.8 compressive strength specified for 15 cm cubes.

   (d) *Central Road Research Institute, New Delhi,* has carried out tests with a view to establish relations between water-cement ratio and compressive strength of concrete using ordinary Portland cements manufactured in the country in accordance with BIS specifications.

As a result of these, it has been considered advisable to give graphs showing the relationship between the compressive strength of concrete mixes with different water-cement ratios and 7 day compressive strength of cement tested in accordance with BIS-Codes. These graphs have been given in Fig. 3.1 as they would be of some assistance in obtaining the water/cement ratio for trial mixes of concrete.

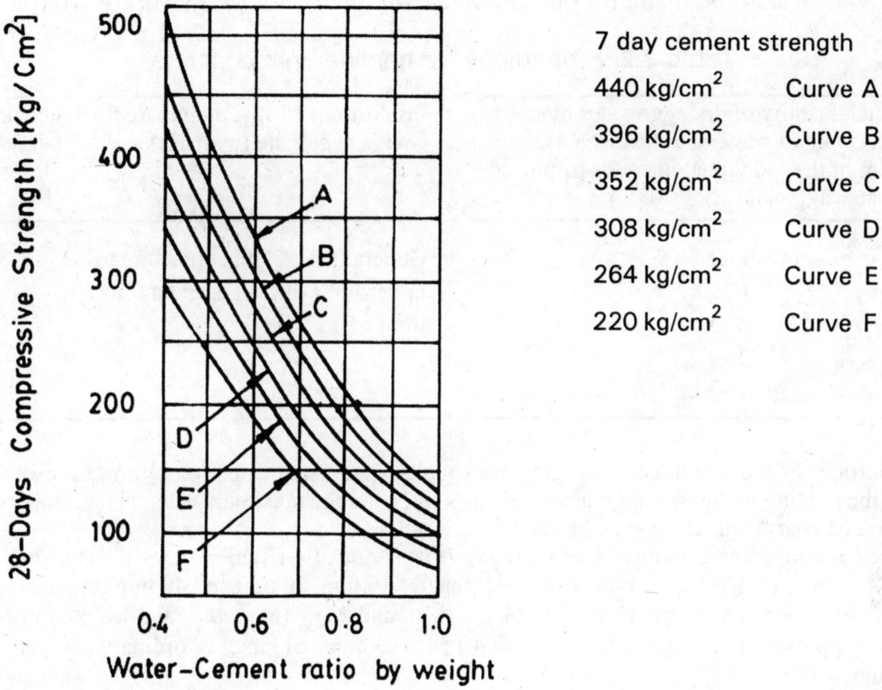

Fig. 3.1. Design curve for cement concrete mixes in relation to 7 days compressive strength of cement (SP:23–1982)

3.3 CONCRETE MIX PROPORTIONING

3.3.1 Methods of proportioning

The determination of the proportions of cement, aggregates and water to attain the required strength shall be made by one of the following :

(a) With preliminary tests by designing the concrete mix; such concrete shall be called 'Design mix concrete', or

(b) Without preliminary tests by adopting nominal concrete mix; such concrete shall be called 'Nominal mix concrete'.

Design mix concrete is preferred to nominal mix. If design mix concrete cannot be used for any reason on the work for grades M20 or lower, nominal mixes may be used with the permission of engineer-in-charge, which, however, is likely to involve a higher cement content.

3.3.2 Nominal mix concrete

Where it is considered not practicable to use design mix concrete of grades M20 or lower the proportions of materials for nominal concrete mixes shall be in accordance with Table 3.2.

In proportioning concrete, the quantity of cement should be determined by weight. The quantities of fine and coarse aggregate may be determined by volume, but these should also preferably be determined by weight. In the latter case the weight should be determined from the volume specified in Table 3.2. and weight per litre of dry aggregate. If fine aggregate is moist and volume batching is adopted, allowance shall be made for bulking in accordance with IS:2386 (Part III).

Table 3.2. Proportions for nominal mix concrete

Grade of concrete	Total quantity of dry aggregates by mass per 50 kg of cement, to be taken as the sum of the individual masses of fine and coarse aggregates, kg (Max.)	Proportions of fine aggregate to coarse aggregate (by mass)	Quantity of water per 50 kg of cement, 50 kg of cement, (Max.) litres
M_5	800	Generally 1 : 2 but subject to an upper limit of 1 : 1½ and a lower limit of 1 : 2½	60
$M_{7.5}$	625		45
M_{10}	480		34
M_{15}	330		32
M_{20}	250		30

Note :

1. The proportions of the fine to coarse aggregates should be adjusted from upper limit to lower limit progressively as the grading of the fine aggregates becomes finer and the maximum size of coarse aggregate becomes larger. Graded coarse aggregate shall be used.
   Example : For an average grading of fine aggregate (that is Zone II of Table 4 of IS:383) the proportion shall be 1 : 1½, 1 : 2 and 1 : 2½ for maximum size of aggregates 10 mm, 20 mm and 40 mm respectively.
2. It may be noted for general guidance that M_{100}, M_{150}, and M_{200} (M_{10}, M_{15} and M_{20}) of ordinary concrete correspond approximately to 1 : 3 : 6, 1 : 2 : 4 and 1 : 1½ : 3 nominal mixes of ordinary concrete currently used in the country.
   The cement content of the mix specified in Table 3.2 for any nominal mix may be increased if the quantity of water in a mix has to be increased to overcome the difficulties of placement and compaction, so that the water-cement ratio as specified is not exceeded.
3. In the case of vibrated concrete, the limit specified may be suitably reduced to avoid segregation.
4. The quantity of water used in the concrete mix for reinforced concrete work should be sufficient, but not more than sufficient to produce a dense concrete of adequate work ability for its purpose, which will surround and properly grip all the reinforcement. Workability of the concrete should be controlled by maintaining a water-cement ratio that is found to give a concrete which is just sufficiently wet to be placed and compacted without difficulty with the means available.

Workability of the concrete should be controlled by direct measurement of water content making allowance for any surface water in the fine and coarse aggregates. The slump-test in accordance with IS:1199 may be used as a guide.

Allowance should be made for surface water present in the aggregate when computing the water content. Surface water shall be determined by one of the field methods described in IS:2368 (Part III). In the absence of exact data, the amount of surface water may be estimated from the values given in Table.

If ordinary concrete made in accordance with the proportions given for a particular grade does not yield the specified strength due to proper qualities of material not being available, such concrete shall be classified as belonging to the appropriate lower grade. Ordinary concrete proportioned for a given grade in accordance with Table 3.2 shall not, however, be placed in a higher grade on the ground that the test strengths are higher than the minimum specified. No interpolation shall be permissible.

3.4 SAMPLING AND STRENGTH TEST OF CONCRETE

3.4.1 General

Samples from fresh concrete shall be taken as per IS:1199 and cubes shall be made, cured and tested at 28 days in accordance with IS:516.

Note : For relatively small and unimportant buildings and works in which quantity of concrete is less than 15 m^3, the strength tests may be waved by the engineer-in-charge at his discretion.

(a) In order to get a relatively quicker idea of the quality of concrete, optional tests on beams for modulus of rupture at 72 ± 2 hours or at 7 days, or compressive strength tests at 7 days may be carried out in addition to 28 day compressive strength tests. For this purpose, the values given in Table 3.3 below may be taken for general guidance in the case of concrete made with Ordinary Portland Cement. In all cases, the 28 days compressive strength shall alone be the criterion for acceptance or rejection of the concrete. If, however, from tests carried out in a particular job over a reasonably long period, it has been established to the satisfaction of the engineer-in-charge that a suitable ratio between 28 days compressive strength and the modulus of rupture at 72 ± 2 hours or 7 days or compressive strength at 7 days may be accepted, the engineer-in-charge may suitably relax the frequency of 28 days compressive strength, provided the expected strength values at the specified early age are consistently met.

Table 3.3. Optional test requirements of concrete

Grade of concrete	*Compressive strength of 15 cm cubes, min, at 7 days N/mm²*	*Modulus of rupture by beam test, min.*	
		at 72 ± 2 h	*at 7 days*
	N/mm²	*N/mm²*	*N/mm²*
M_{10}	7.0	1.2	1.7
M_{15}	10.0	1.5	2.1
M_{20}	13.5	1.7	2.4
M_{25}	17.0	1.9	2.7
M_{30}	20.0	2.1	3.0
M_{35}	23.5	2.3	3.2
M_{40}	27.0	2.5	3.4

3.4.2 Frequency of sampling

(a) *Sampling procedure* : A random sampling procedure shall be adopted to ensure that each concrete batch shall have a reasonable chance of being tested, that is, the sampling should be spread over the entire period of concreting and cover all mixing units.

(b) *Frequency* : The minimum frequency of sampling of concrete of each grade shall be in accordance with Table 3.4.

Table 3.4

Quantity of concrete in the work, m^3	*Number of samples*	
1-5	1	
6-15	2	
16-30	3	
31-50	4	
51 and above	4	plus one additional sample for each additional 50 m^3 or part thereof

Note : At least one sample shall be taken from each shift. Where concrete is produced at continuous production unit, such as ready-mixed concrete plant, frequency of sampling may be agreed upon mutually by suppliers and purchases.

3.4.3 Test specimen

Three test specimens shall be made for each sample for testing at 28 days. Additional samples may be required for various purposes such as to determine the strength of concrete at 7 days or at the time of striking the form work, or to determine the duration of curing or to check the testing error. Additional samples may also be required for testing cubes cured by accelerated methods as described in IS:9103. The specimen shall be tested as described in IS:516.

3.4.4 Test results of sample

The test results of the sample shall be the average of the strength of three specimens. The individual variation should not be more than $\pm$ 15 per cent of the average. If more, the test results of the sample are invalid.

3.4.5 Standard deviation

The standard deviation for each grade of concrete shall be calculately separately.

3.4.5.1 Standard deviation based on test results strength of sample

(a) *Number of test results of samples* : The total number of test strength of samples required to constitute an acceptable record for calculation of standard deviation shall not be less than 30. Attempts should be made to obtain the 30 samples, as early as possible, when a mix is used for the first time.

(b) *In case of significant changes in concrete* : When significant changes are made in the production of concrete batches (for example changes in the materials used, mix design, equipment or technical control), the standard deviation value shall be separately calculated for such batches of concrete.

(c) *Standard deviation to be brought upto date* : The calculation of the standard deviation shall be brought upto date after every change of mix design.

3.4.5.2 *Determination of standard deviation*

(a) Concrete of each grade shall be analysed separately to determine its standard deviation.
(b) The standard deviation of concrete of a given grade shall be calculated using the following formula from the results of individual tests of concrete of that grade obtained as specified in 3.4.4.

Estimated standard deviation $s = \sqrt{\dfrac{\Sigma D^2}{n-1}}$

where :

D - deviation of the individual test strength from the average strength of n samples, and

n - number of sample test results.

(c) When significant changes are made in the production of concrete batches (for example changes in the materials used, mix design, equipment or technical control), the standard deviation value shall be separately calculated for such batches of concrete.

3.4.5.3 *Assumed standard deviation*

Where sufficient test results for a particular grade of concrete are not available, the value of standard deviation given in Table 3.5 may be assumed for design of mix in the first instance. As soon as the result of samples are available actual calculated standard deviation shall be used and the mix designed properly.

Table 3.5. Assumed standard deviation

Grade of concrete	Assumed standard deviation N/mm^2
M_{10}	3.5
M_{15}	3.5
M_{20}	4.0
M_{25}	4.0
M_{30}	5.0
M_{25}	5.0
M_{40}	5.0
M_{45}	5.0
M_{50}	5.0

Note : The above values correspond to the site control having proper storage of cement, weigh batching of all materials, controlled addition of water; regular checking of all materials, aggregate gradings and moisture content; and periodical checking of workability and strength. Where there is deviation from the above, the values given in the above table shell be increased by $1 N/mm^2$.

However, when adequate past records for a similar grade exist and justify to the designer a value of standard deviation different from that shown in the table above it shall be permissible to use that value.

3.5 ACCEPTANCE CRITERIA

3.5.1 Compressive strength

The concrete shall be deemed to comply with the strength requirements when both the following conditions are met :

(a) The mean strength determined from any group of four consecutive test results complies with the appropriate limits in Col. 2 of Table 3.6.

(b) Any individual test result complies with the appropriate limits in Col. 3 of Table 3.6.

Table 3.6. Characteristic compressive strength compliance requirement

Specified grade	*Mean of the Group of 4 Non-overlapping consecutive test results in N/mm²*	*Individual test results in N/mm²*
(1)	(2)	(3)
M 15	$\geq f_{ck}$ + 0.825 × established standard deviation (rounded off to nearest 0.5 N/mm²) or f_{ck} + 3N/mm², whichever is greater	$\geq f_{ck}$ -3 N/mm²
M 20 or above	$\geq f_{ck}$ + 0.825 × established standard deviation (rounded off to nearest 0.5 N/mm²) or f_{ck} + 4 N/mm², whichever is greater	$\geq f_{ck}$-4 N/mm²

Note : In the absence of established value of standard deviation, the values given in Table 3.5 may be assumed, and attempt should be made to obtain results of 30 samples as early as possible to establish the value of standard deviation.

3.5.2 Flexural strength

When both the following conditions are met, the concrete complies with the specified flexural strength.

(a) The mean strength determined from any group of four consecutive test results exceeds the specified characteristic strength by at least 0.3 N/mm².

(b) The strength determined from any test result is not less than the specified characteristic strength less 0.3 N/mm².

3.5.3 Quantity of concrete represented by strength test results

The quantity of concrete represented by a group of four consecutive test results shall include the batches from which the first and last samples were taken together with all intervening batches.

For the individual test result requirements given in column 2 of Table 3.6 or in item (b) of 3.5.2, only the particular batch from which the sample was taken shall be at risk.

Where the mean rate of sampling is not specified the maximum quantity of concrete that four consecutive test results represent shall be limited to 60 m³.

3.5.4 If the concrete is deemed not to comply persuant to 3.5.3, the structural adequacy of the parts affected shall be investigated and any consequential action as needed shall be taken.

3.5.5 Concrete of each grade shall be assessed separately.

3.5.6 Concrete is liable to be rejected if it is porous or honey-combed, its placing has been interrupted without providing a proper construction joint, the reinforcement has been displaced beyond the tolerances specified, or construction tolerances have not been met. However, the hardened concrete may be accepted after carrying out suitable remedial measures to the satisfaction of the engineer-in-charge.

3.6 SUGGESTED PROPORTIONS OF FINE AGGREGATE TO COARSE AGGREGATE

As stated in note 1 of Table 3.2 for 'Concrete Mix Proportion', Table 3.7 provides guidelines for adopting FA : CA ratio to be adopted for the first trial mix.

Table 3.7. Proportions of fine aggregate to coarse aggregate for ordinary concrete mix

Nominal size of coarse aggregate	Zones of sand			
	Zone I	Zone II	Zone III	Zone IV
10 mm	1 : 1	1 : 1½	1 : 2	1 : 2½
20 mm	1 : 1½	1 : 2	1 : 2½	1 : 3
40 mm	1 : 2	1 : 2½	1 : 3	—

3.7 SIZES AND DIMENSIONS OF CONTAINERS FOR VOLUME BATCHING OF ORDINARY CONCRETE MIX (Concrete Handbook, CAI, Bombay)

Table 3.8. Size and dimensions of containers

Capacities in litres	Inside measurements		
	length (cm)	breadth (cm)	height (cm)
25	25	25	40
30	25	25	48
35	27	27	48
40	29	29	48
45	30	30	50
50	31	31	52

3.8 SUGGESTED WORKABILITY RANGES

Table 3.9. Suggested ranges of workability of concrete for different placing conditions (SP:23–1982)

Placing conditions	Degree of workability	Slump (mm)
Blinding concrete; Shallow sections; Pavements using pavers	Very low	0.75–0.80 C.F.
Mass concrete; Lightly reinforced sections in slabs, beams, walls, columns; Floors; Hand placed pavements; Canal lining; Strip footing	Low	25–75
Heavily reinforced sections in slabs, beams, walls, columns; Slipform work; Pumped concrete	Medium	50–100 75–100
Trench fill In situ piling	High	100–150
Tremie concrete	Very high	Measurement should be made by determination of flow

Note : For most of the placing conditions, internal vibrators (needle vibrators) are suitable. The diameter of the needle shall be determined based on the density and spacing of reinforcement bars and thickness of sections. For tremie concrete, vibrators are not required to be used.

3.9 TYPICAL SLUMP RANGES PRESCRIBED BY PORTLAND CEMENT ASSOCIATION, ILLINOIS

Table 3.10. Slump ranges for various types of constructions

Types of construction	*Slump*	
	Max. (mm)	*Min. (mm)*
Reinforced foundation walls and footings	100	50
Unreinforced footing caissons and substructure walls	75	25
Reinforced slabs, beams and columns	125	50
Building columns	125	75
Bridge decks	75	50
Pavements	50	25
Side-walls, drive ways and slabs on ground	100	50
Heavy mass construction	50	25

3.10 DEGREE OF WORKABILITY FOR VARIOUS REQUIREMENTS AS SUGGESTED BY CAI, MUMBAI (Notes on design of concrete mixes by CAI, Mumbai)

Table 3.11. Degree of workability for various requirements

Requirement	*Slump (mm)*
1. Vibrated concrete in walls or other large sections	0–25
2. Mass concrete foundations without vibrations, simple reinforced section with vibration	25–50
3. Normal reinforced work without vibrations and heavily reinforced section with vibration	50–100
4. Sections with congested reinforcements not normally suitable for vibrations	100–175

CHAPTER 4

Method of Sampling Cement (Adopted from IS:3515)

4.1 GENERAL

Tests on various properties of cement, aggregate and concrete are performed on the samples taken from a heap. The result of the tests strictly speaking are applicable to the samples tested. But since the aim of all such tests is to ascertain qualities of the bulk of cement, aggregate and concrete, it is imperative that the sample collected represent the typical of average properties of material to be tested. Such sample is said to be the representative sample. Much care and systematic procedure is required to be followed. These methods are now standardized by the ISI.

4.2 SAMPLING OF CEMENT

1. Sampling instrument should be clean and dry when used.
2. Precautions should be taken to protect the sample, the material being sampled, sampling implements and containers for samples from contamination.
3. Sample containers should be of such a size that they are almost completely filled by the sample.
4. Sample container shall be sealed air tight after filling and marked with full particular of the material and the date of sampling.
5. Samples shall be stored in such a manner that the temperature of the material does not vary unduly from normal temperature.

4.3 TERMINOLOGY

Quantity of cement offered by the manufacturing unit at one time is called a *lot*. This *lot* is normally divided into *sublots* depending upon its total weight as per Table 4.1. Quantity of cement taken by a single operation is called *increment*. Aggregates of all these increments from the same sublot is called *gross sample*. When these gross samples of each sublot are reduced by suitable procedures for laboratory testing, they are called *laboratory samples*. Equal quantities of cement from each laboratory sample representing sublots, are mixed to form composite samples.

Table 4.1. Number of sublots into which a lot is to be divided

Weight of lot (tonnes)	Number of sublots
Up to 1002	
101 to 200	3
201 to 300	4
301 to 500	5
501 to 800	6
801 to 1300	7
1301 and above	8

4.4 SAMPLING

For sampling of cement from conveyor, bulk-storage and ship's hold and wagons, the reader is referred to IS : 3535.

Only the method of sampling from bags is reproduced below since a supervisor is often concerned with such type of sampling :

(i) Divide the entire lot into sublots as per Table 4.1. Sublots should consist of equal number to bags.

(ii) From each sample at least two per cent bags should be sampled.

(iii) These bags are to be chosen at random. To ensure randomness following procedure should be followed :

(a) If there are 'N' number of bags in the sublots and 'n' is the number of bags to be selected, then calculate r = N/n.

(b) Starting from any bag, count the bags in one order as 1, 2, 3, upto r. Every 'r'th bag so counted should be removed.

(c) For example, if a lot consists of 3000 bags i.e. 150 tonnes, divide this lot into 3 sublots i.e. 1000 bags each. Thus N = 1000. Since at least, 2 per cent bags are to be sampled, then n = 20 bags. Therefore, r = 1000/20 = 50. Thus starting from any one point, every 50th bag is to be removed.

Diagrammatic explanation of obtaining the suitable samples of cement for testing is given in Fig. 4.1.

(iv) Increments are to be taken by inserting a tube sampler as specified in IS : 3535. The material so collected from each bag is 0.75 kg.

(v) Material so collected should be placed directly into moisture proof, air tight containers to avoid, moisture absorption and aeration of the sample. Cans, plastic bags or moisture-proof multiple wall paper bags may be used. These containers shall be filled immediately as sealed. The containers shall carry suitable identification marks so that they can be referred back to the particular sublots which the gross sample have been taken.

Gross-samples thus obtained have to be reduced to laboratory samples by quartering or using riffler box. Quartering should be done as shown in Fig. 4.1.

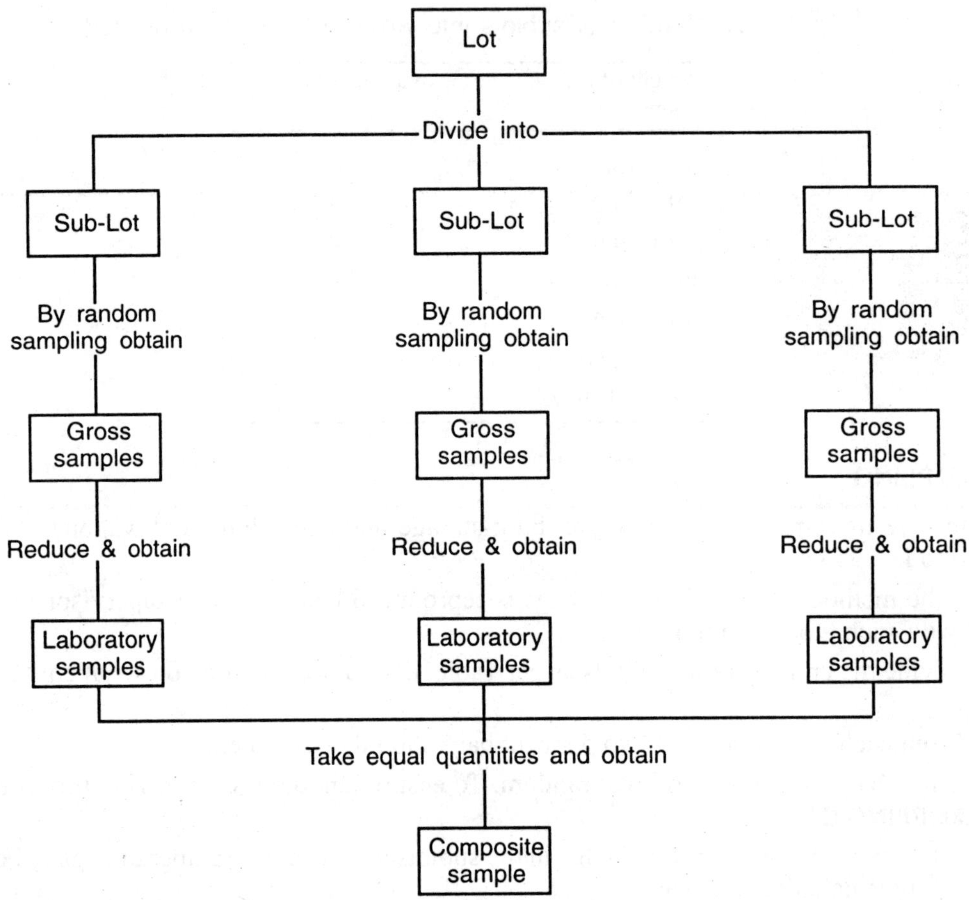

Fig. 4.1. Diagrammatic sketch for sampling of cement.

The material should be scooped into cone-shaped pile. Care shall be taken to drop each scoopful exactly over the same spot, as otherwise the central-axis of the cone will be slackened. After the cone is formed, it shall be flattened by pressing the top of the cone with smooth surface of the scoop. Then the cone is cut into the quarters by two lines which intersect at right angles at the centre of the cone. The reduction is achieved by rejecting any two diagonally opposite quarters.

Reduction should be continued till about 7 kg of material required for laboratory sample are obtained.

(vi) Equal quantities of material shall be taken from each laboratory sample representing the sublot and mixed together to constitute composite sample representing the lot as a whole. The weight of the composite sample shall be about 7 kg.

(vii) All laboratory samples and composite samples should be sealed in an air tight moisture proof container.

CHAPTER 5

Method of Sampling Aggregate

5.1 GENERAL

Correct sampling in case of aggregates is all the more important, since an engineer has to choose from the naturally available material and its manufacture cannot be subjected to the same strict control as that of cement.

5.2 SAMPLING OF CEMENT

The words lot, sublots, gross-sample, laboratory sample, composite samples have the same meaning at that explained under 'sampling of cement'.

Aggregate sampling from conveyor, wagons, trucks and boats are necessary and for reference to the methods of sampling, the student is referred to IS : 2439.

The method of sampling from stacks on site only is described here since it is of immediate importance to the site supervisor.

5.3 PROCEDURE

(i) Divide the lot in sublots as per Table 5.1.

(ii) Minimum weight of gross samples are the minimum number of increments should be as shown in Table 5.2.

(iii) Increments should be drawn with the help of scoop (Fig. 5.1) of the sizes mentioned in Table 5.3.

(iv) Sampling of aggregate from the stacks should be carried out, as far as possible, during the making (or breaking) of the stack.

(v) Number of increments should be equally distributed over the sub-lot (Fig. 5.2).

(vi) When it becomes necessary to sample a stationary stack, trench sampling method may be used. This method is applicable for the stacks upto a maximum height of 1.5 m only. The method is as follows :

Table 5.1. Number of sublot into which a lot is to be divided

Size of lot (m^3)	*Number of sublots*
101 to 500	3
501 to 1500	5
1501 to 5000	7

Note : In case the lot of aggregates is less than 100 m^3, the sampling shall be subjected to agreement between the purchaser and the supplier.

Table 5.2. Minimum weight of gross sample and minimum number of increment

Max size present in substantial proportions	*Wt. of gross sample min*	*Weight of increment*	*Number of increments min*
Coarse and all-in-aggregates (more than 20 mm aggregate size)	300	10	30
Coarse and all-in-aggregates (up to 20 mm nominal size)	200	5	40
Fine aggregates	100	2	50

Table 5.3. Dimensions of Scoop

Capacity of scoop	*Dimensions of the scoop*			
	A	*B*	*C*	*D*
kg	mm	mm	mm	mm
10	250	250	120	200
5	200	200	90	175
2	150	150	75	125

Along a randomly chosen line on the aggregate surface or the sublot, a trench shall be dug right down to the ground level (Fig. 5.2, 5.3) leaving about half a meter of walking space at the ground level. From the trenches so dug, the required number of increments shall be collected with the help of a suitable sampling scoop at various points randomly spread over the two exposed side of the trenches. In case of large stacks, in addition to the trench, the side of the piles may also be opened to expose the aggregate down to the bottom at places where the trench does not expose the aggregate inside.

(vii) Each gross sample should be reduced separately. Reduction may be done with the help of Riffle divider, or by coning and quartering method.

In riffling a sample is discharged over the full width of the riffler with the help of a specially designed scoop. The two halves separated by the riffler is collected at the bottom by two

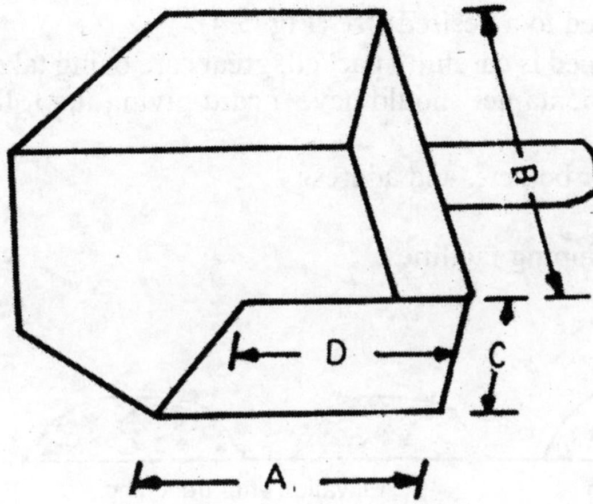

Fig. 5.1. Sampling scoop.

Fig. 5.2. Collecting aggregate from a heap.

boxes on each side of the chute. One half is discarded and the riffling of the other half is repeated till the sample is reduced to a desired size (Fig. 5.4).

(viii) Each laboratory sample so obtained is carefully packed, great care being taken to prevent the loss of any fine material. Each container should have a card giving the following information:

(a) Name of the quarry pit, river bed etc. and address.

(b) Proposed use of material.

(c) Geographic location and shipping facilities.

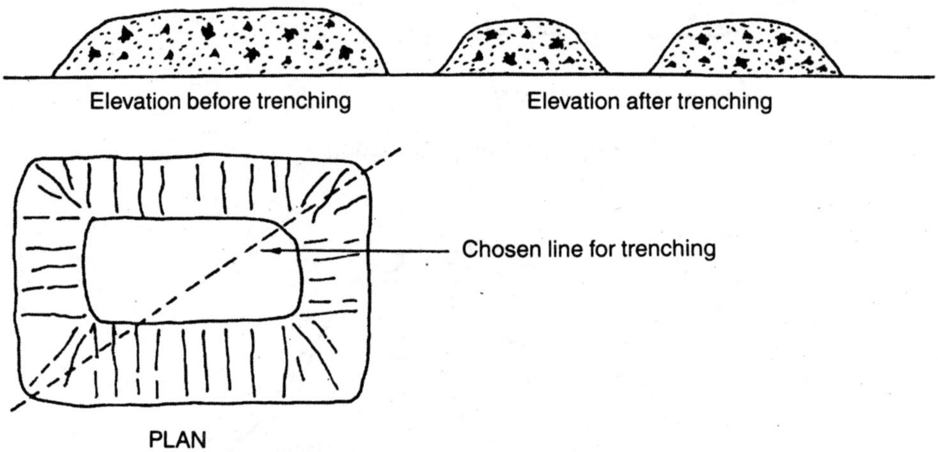

Fig. 5.3a. Trench sampling method.

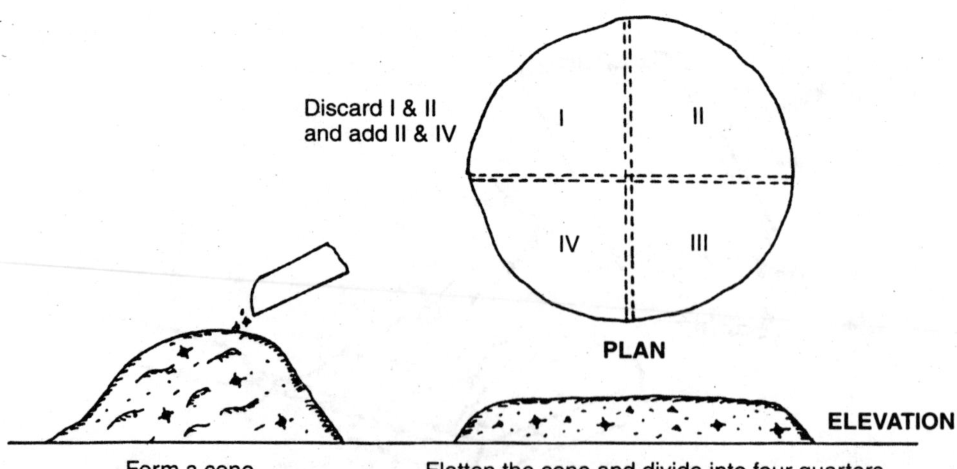

Fig. 5.3b. Reduction of samples by coning.

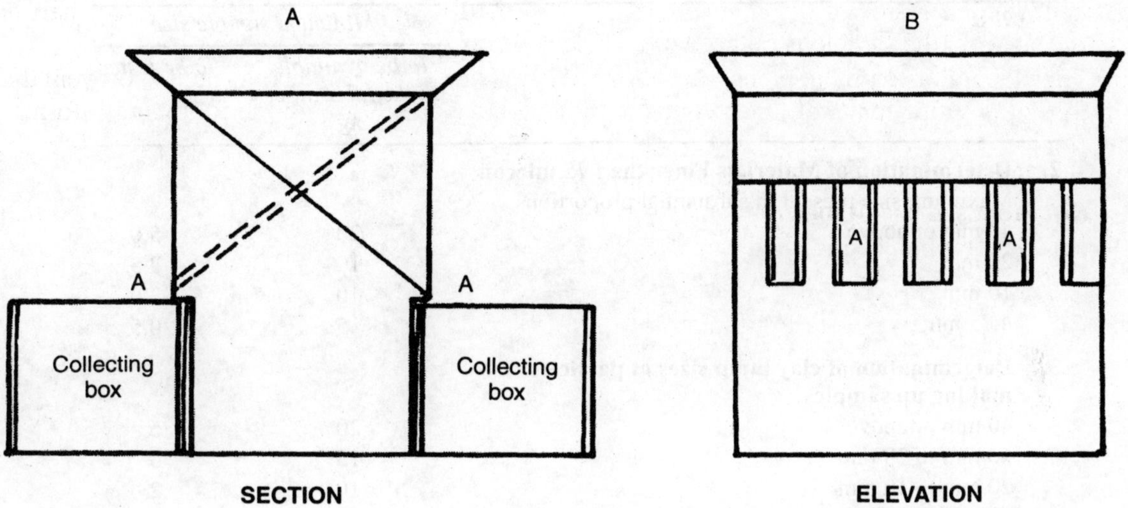

Fig. 5.4. Sample splitter riffler box.

5.4 SIZES OF SAMPLES REQUIRED FOR VARIOUS TESTS IN ACCORDANCE WITH IS : 2386 (Adopted from Concrete Handbook, CAI, Bombay)

Table 5.4. Sizes of samples required for various tests

Test	*Minimum sample size*	
	To be available in the laboratory kg	*To be tested kg*
1. **Sieve Analysis**		
Maximum size present in substantial proportions		
63 mm	150	50
50 mm	150	35
40 mm to 31.5 mm	50	15
25 mm	15	5
20 mm to 16 mm	10	2
12.5 mm	5	1
10 mm	2	0.5
6.3 mm	2	0.2
4.75 mm	2	0.2
2.36 mm	2	0.1

(Contd.)

	Test	*Minimum sample size*	
		To be available in the laboratory kg	*To be tested kg*
2.	**Determination of Materials Finer than 75 micron**		
	Maximum size present in substantial proportions		
	40 mm or above	20	5.0
	20 mm	10	2.5
	10 mm	10	2.0
	4.75 mm	5	0.5
3.	**Determination of clay lump sizes of particles making up samples**		
	40 mm or above	20	5
	40 mm to 20 mm	15	3
	20 mm to 10 mm	10	2
	10 mm to 4.75 mm	5	1
4.	**Determination of clay, fine silt and fine dust (by the Sedimentation method)**		
	Maximum size present in substantial proportions		
	63 mm to 25 mm	30	6
	20 mm to 12.5 mm	20	1
	10 mm to 6.3 mm	10	0.5
	4.75 mm and smaller	10	0.3
5.	**Determination of Light Weight Pieces (Coal and Lignite)**		
	Weight pieces (coal and lignite)		
	80 mm	50	10
	40 mm	25	5
	20 mm	15	3
	6.3 mm	5	0.2
6.	**Determination of Soft Particles**		
	Maximum size of aggregate		
	40 mm to 30 mm	60	12
	30 mm to 25 mm	30	4.5
	25 mm to 20 mm	10	1.5
	20 mm to 12.5 mm	5	0.6
	12.5 mm to 10 mm	5	0.2
7.	**Estimation of organic Impurities**	5	0.5
8.	**Determination of Specific Gravity and Water Absorption**		
	Aggregate larger than		
	10 mm	10	2
	10 mm to 4.75 mm	5	1
	Smaller than 4.75 mm	3	0.5

(Contd.)

Test	Minimum sample size	
	To be available in the laboratory kg	*To be tested kg*
9. Determination of Bulk Density and Voids		
Max. size of aggregate		
Over 40 mm	150	60
40 mm to 4.75 mm	100	30
Less than 4.75 mm	20	6
10. Determination of Aggregate Crushing Value		
Crushing value	30	7
11. Determination of Ten per cent fines value	30	7
12. Determination of Aggregate Impact value	5	1
13. Determination of Aggregate Abrasion value by use of Deval machine	30	6
14. Determination of Aggregate Abrasion value by use of Los Angeles Machine	300	50
15. Determination of Fineness		
Coarse aggregate	60	10
Fine aggregate	10	1

CHAPTER 6

Method of Sampling Concrete

6.1 GENERAL

Criteria for the acceptance of concrete is mainly based on the strength characteristics of the concrete, which are obtained by conducting compressive tests on the cubes or cylinder. The validity of the test results depends to a large extent on the care with which representative samples of concrete are collected and cube or cylinders cast.

The method explained below specifies the procedure to be followed in the field for obtaining representative samples of fresh concrete directly from the mixer or from the concrete at the time and place of deposit. This is adopted from IS : 1199 and IS : 516.

6.2 SAMPLING OF CEMENT

Sample from a batch should not be less than 0.02 m^3 in volume i.e. 20 litres. This should compose of portions taken from different points in a batch.

6.3 PROCEDURE

From mixer : At least three approximately equal sample increments totalling 0.02 m^3 (20 litres) shall be taken from a batch *during its discharge*. Each sample increments is to be collected by passing clean dry receptacle across a stream of concrete. Receptacle should be such as to avoid segregation. Buckets of 7 to 10 litres capacity of G.I. will serve the purpose.

Where three increments are taken, sample should be taken at about a time, when one quarter, one half and three quarters of the concrete have been discharged from the mixer.

6.4 FROM CONCRETE AT THE TIME AND PLACE OF DEPOSITION

Sample shall be taken while a batch of concrete is being prepared or immediately after it has been discharged on site. Sample shall be collected at not less than five well distributed position, avoiding edge of the mass where segregation may have occurred.

6.5 MIXING THE COMPOSITE SAMPLE

The composite sample obtained by either of the methods mentioned above shall be mixed on a non-absorbent base either with a shovel or trowel to ensure uniformity. These samples can then be used for conducting :

(a) slump test,
(b) compacting factor test,
(c) analysis of freshly mixed concrete,
(d) preparation of cube for strength test.

6.6 CURING OF FIELD SAMPLE FOR WORKS-CUBE TEST

The concrete from the composite sample mentioned above is cast in the cube or beam or cylinder mould in a way prescribed by the Indian Standard Institution and then should be stored and cured properly.

The specimen should be stored on site at a place free from vibration, under damp matting or sacks for 24 hours ± ½ hour from the time of adding water. The temperature should be between 22°C to 32°C. After 24 hours they shall be *marked for identification*, removed from the mould and kept in clean cool water till they are transported to laboratory for testing. Transporting should be done by packing them well in damp sand or damp sacks. After the arrival in laboratory, they should be kept in water till the time of test.

SAMPLING OF MATERIALS FOR CONCRETE FOR PRELIMINARY TESTING IN LABORATORY

6.7 SAMPLING OF CEMENT

Test samples made up of small portions should be taken from each of a number of bags on site. Further details may be seen in Chapter 4 - "Methods of Sampling Cement".

6.8 AGGREGATE

About 600 kg of aggregate should be collected from the site as per the procedure mentioned under 'Sampling of Aggregate'. Out of this lot, collect suitable quantity of riffling or quartering.

6.9 PREPARATION OF MATERIALS

(a) All material should be brought to the room temperature preferably 27°C ± 3°C before commencing tests.
(b) Cement on arrival at the laboratory, shall be thoroughly mixed to ensure uniformity in material. Cement shall be stored in a dry place, preferably in air-tight metal container.
(c) Aggregate should be in air-dry condition and should be separated in fine and coarse fractions. 4.75 mm sieve should normally be used for such separation. These may be recombined in specified proportions according to test specifications. Where special grading is mentioned, the coarse and fine fractions should be separated into different sizes.

CHAPTER 7

Frequency of Control Tests for Quality Concrete

7.1 GENERAL

Since concrete is composed of materials whose properties vary from time time and place to place, there is always large variations between the minimum strength and the mean strength of the concrete mix. Smaller this difference, lesser is the quantity of cement to be used. Effort should, therefore, be made to control the variations in the properties of the mix ingredients and also control the accuracy of all operations which effect the strength and consistency of concrete i.e. batching, mixing, placing, curing and testing.

This chapter deals with the tests to control the variations in the properties of mix ingredients.

7.2 CONTROL ON CEMENT

Cement is usually manufactured under controlled conditions. Tests are rarely conducted on site but they must be conducted in laboratory under controlled conditions.

Only when the cement is stored at one place for a long time or is adulterated during transit, it is likely to deteriorate in quality. Field test as explained in 'experiment No. 6' should be carried out to detect adulteration. If there is doubt, the cement samples should be sent to laboratory for :

(a) fineness test,
(b) setting time test,
(c) soundness test, and
(d) strength test.

Whenever high strength concrete is being manufactured, the minimum strength is specified. The compression test on cement should invariably be carried out.

Initially cement should be tested once for each source of supply and subsequently after every two months.

7.3 QUALITY CONTROL ON WATER

Normally any water fit for drinking is suitable for concrete. But water from ponds rivers, and canals are of doubtful purity and may contain organic matter having serious retarding effect on hydration. Similarly water containing sugar, excess of alkali, acid and salt shall not be used.

The best way to test the suitability is to conduct setting time test and compression test on cement by using the water under test and comparing its results with corresponding values obtained with the same cement mixed with distilled water.

In case of doubt regarding development of strength, the suitability of water for making concrete shall be ascertained by the compressive strength and initial setting time tests specified below :

The sample of water taken for testing shall represent the water proposed to be used for concreting, due account being paid to seasonal variation. The sample shall not receive any treatment before testing other than that envisaged in the regular supply of water proposed for use in concrete. The sample shall be stored in a clean container previously rinsed out with similar water.

Average 28 days compressive strength of atleast three 15 cm concrete cubes prepared with water proposed to be used shall not be less than 90 per cent of the average of strength of three similar concrete cubes prepared with distilled water. The cubes shall be prepared, cured and tested in accordance with the requirements of IS : 516.

The initial setting time test of block made with the appropriate cement and the water proposed to be used shall not be less than 30 minutes and shall not differ by $\pm$ 30 minutes from the initial setting time of control test block prepared with the same cement and distilled water. The test blocks shall be prepared and tested in accordance with the requirements of IS : 4031.

The pH value of water shall generally be not less than 6.

7.4 QUALITY CONTROL ON AGGREGATE

For approval of each source of supply, aggregates are required to be tested specially for specific gravity, water absorption, deleterious material and organic impurities.

Subsequently tests are required to be conducted once or twice daily for grading and moisture contents.

Normally maximum moisture is found to be present in lower layers of the heap and upper layers are comparatively dry. When the sand is being drawn from this layer, the moisture content is required to be determined. After few tests, moisture content can be judged by eye and feel. Subsequently moisture test should be carried out when the change in moisture is readily apparent.

Every time the moisture content is determined, bulkage test on sand is also required to be carried out if the batching is done by volume.

Normally for every sand, it is advisable to make a field test to determine whether the sand containers silt in appreciable quantities. This test will give clue whether the sand should be washed or not.

Test for grading of aggregate is required to be made once daily or when the new material arrive or when it is taken from different heaps. This helps to maintain uniformity of grading. Aggregate containing too much or too little of one size should be avoided. Some specifications insist on at least 10 per cent of finer material passing through 150 mic sieve and retained on 75 mic sieve. Sieve analysis help to determine this percentage.

Moisture content varies from layer to layer in aggregate. Effort must, therefore, be made to stack aggregate in such a way as to ensure uniformity of moisture. Place aggregates in layers not thicker than those resulting from lorry dumped in some place. Pyramids should be avoided. Stock-piles should be spread in as large areas as possible and the height of the stock-pile should not exceed 1.25 m to 1.5 m. Aggregate should be left standing 24 hours before use. Bottom layer of about 15 cm to 30 cm should not be used for concrete, if it is moist.

7.5 QUALITY CONTROL OF CONCRETE

Slump test should be conducted at the start of each operation and subsequently when the change in the moisture content in the aggregate is apparent or there is the uniformity of grading and change in shape. All these changes cause the change in appearance of the concrete and whenever such change is apparent, slump test should be carried out.

At least three tests for workability should be conducted daily for 25-30 m^3 of concrete.

For strength test, acceptance criteria as specified by IS : 456 as given earlier should be strictly adhered to.

CHAPTER 8

Tests on Cement

EXPERIMENT 1

DETERMINATION OF WATER FOR CEMENT PASTE OF NORMAL CONSISTENCY

Statement of problem

For a given sample of cement, determine the percentage of water for normal consistency.

Theory

Cement paste of normal consistency is defined as percentage of water by weight of cement which produces a consistency which permits a plunger of 10 mm dia. to penetrate up to a depth of 5 mm to 7 mm above the bottom of Vicat's Mould.

Before performing the test for initial setting time, final setting time, compressive strength, tensile strength and soundness of cement etc., it is necessary to fix the quantity of water to be mixed to prepare a paste of cement of standard consistency in each case. The quantity of water to be added in each of the above mentioned experiment bears a definite relation with the percentage of water for standard consistency. This experiment is intended to find out for a given cement, the quantity of water to be mixed to give a paste of standard consistency.

Percentage of water in the cement paste 'P' for standard consistency will vary from cement to cement and from batch to batch of the same cement, and the quantities of water used in the other tests will vary accordingly. Following are the quantities of water required for various tests :

(i) Quantity of water for setting time test expressed as percentage of weight of cement = 0.85 P

(ii) Quantity of water for soundness test expressed as percentage of weight of cement

(a) Le Chatelier method = 0.78 P

(b) Autoclave method = P

(iii) Quantity of water for compressive strength on 1 : 3 cement and standard sand mortar expressed as percentage of weight of dry cement and aggregate = P/4 + 3.0 where P is the percentage of water for standard consistency.

Apparatus (Fig. 8.1)

- Vicat's needle apparatus with plunger of 10 mm dia and 50 mm length, weighing 300 g and Vicat's mould.

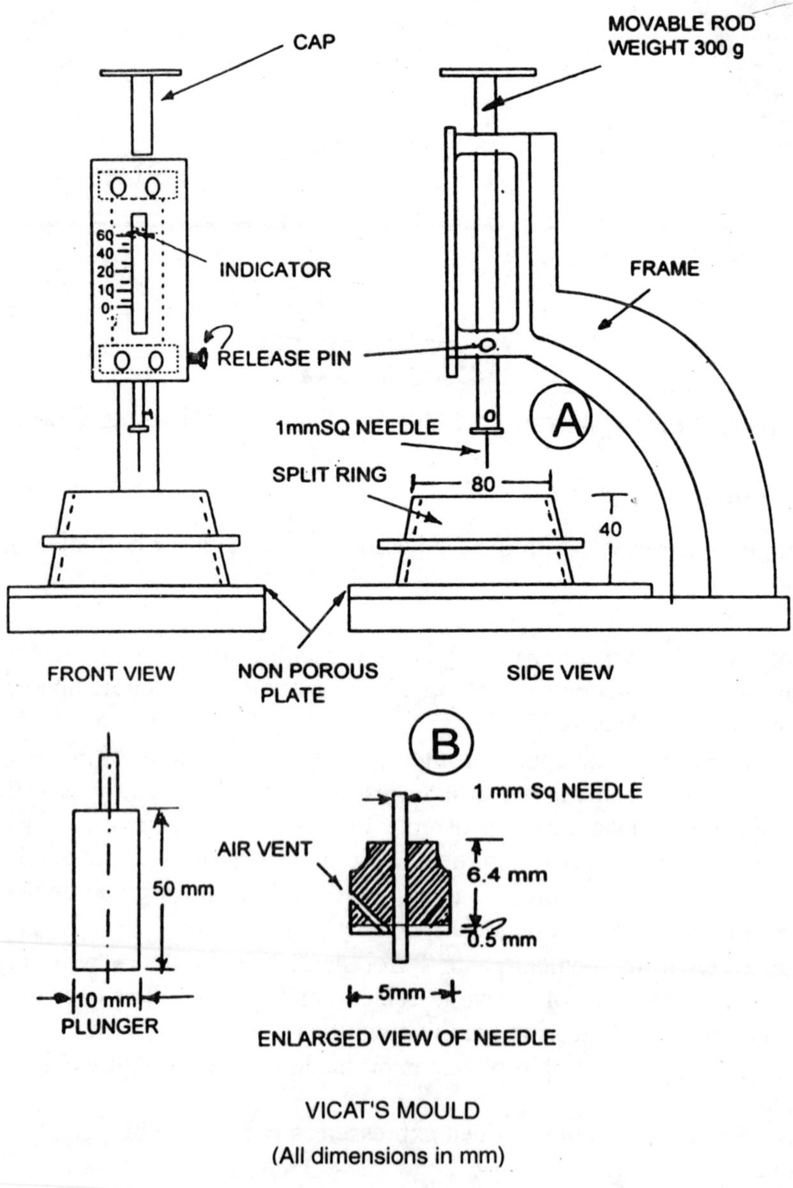

Fig. 8.1. Vicat's mould

- Balance with weights capacity - 1 kg
- Trowel
- Enamel trough
- Standard spatula
- Thermometer (range 50°C)
- Stop watch
- Non-porous plate

Procedure

(a) Weigh about 400 gms of cement accurately and place it in the enamel trough.
(b) To start with, add about 25% of clean water and mix it by means of a spatula. Care should be taken that the time of gauging is not less than 3 minutes and not more than 5 minutes. The gauging time shall be counted from the time of adding water to the dry cement until commencing to fill the mould.
(c) Fill the Vicat mould with this paste, the mould resting on non-porous glass plate.
(d) Make the surface of the cement paste in level with the top of the mould with a trowel weighing 210 g. The mould should be slightly shaken to expel the air.
(e) Place this mould together with the non-porous glass plate under the rod bearing the plunger. Adjust the indicator to show 0-0 reading when it touches the surface of test block.
(f) Release the plunger quickly, allowing it to sink into the paste.
(g) Prepare trial pastes with varying percentages of water and test as described above until the needle penetrates 5 mm to 7 mm above the bottom of the mould.
(h) Express this amount of water as a percentage by weight of the dry cement.
(i) Room temperature at the time of testing shall be 27 $\pm$ 2°C.

Precautions

(a) Clean appliances should be used for gauging.
(b) The temperature of cement and water and that of the test room at the time of test should be from 27 $\pm$ 2°C.
(c) In filling the mould, the operators hand and the blade of the gauging trowel alone be used.

Observation table

1. Details of sample

(a) Type of cement ....
(b) Name of manufacturer ....
(c) Delivered by ....
(d) Date of delivery ....
(e) Delivery receipt number ....
(f) Sample taken from ....
(g) Packing; air tight/not air tight ....
(h) Samples taken by ....

2. Standard consistency test

(a) Date of test

(b) Temperature of testing : (i) Cement .... °C

(ii) Water .... °C

(iii) Room .... °C

(c) Quantity of cement used (C) .... gm

S.No.	*Quantity of water added*	*Time of gauging minutes*	*Peetration from bottom of the mould mm*
1			
2			
3			
4			
5			

Conclusion

Percentage of water required to prepare a cement paste of standard consistency :

$$P = \frac{W}{C} \times 100 =$$

Questions for discussion

1. What is the normal consistency of cement?
2. How does fineness of cement affects the normal consistency of cement?
3. What is the effect of normal consistency of cement on all other physical tests?

EXPERIMENT 2

DETERMINATION OF INITIAL AND FINAL SETTING TIME OF CEMENT

Statement of problem

Given a sample of cement, determine the intial and final setting time of the sample.

Theory

When water is mixed with cement to form a paste, reaction starts. In its pure form, the finely ground cement is extremely sensitive to water. Out of the three main compounds viz. C_3A, C_3S and C_2S, the C_3A reacts quickly with water to produce jelly-like compound which starts solidifying. This action of changing from a fluid state to a solid state is called 'Setting' and should not be confused with 'Hardening'.

During the next stage of hydration, cement paste starts hardening owing to the reaction of C_3S and C_2S and the paste gains strength. In the first few minutes, the setting action is more predominant and after some time hardening action becomes rapid.

In practice, such solidifying action or loss of plasticity is required to be delayed, because some time is needed for mixing transporting and placing of concrete into final position before the mix looses its plasticity due to setting action.

It is usually specified that the plastic concrete should be placed and consolidated before initial set has occurred, it should not then be disturbed until concrete has hardened. This initial setting time should not be too small and therefore, the standards specify minimum initial setting time.

Once initial stiffening of concrete has taken place it is desirable that it should harden or gain strength as rapidly as possible, so that there is minimum of delay before shuttering can be removed and risk of frost-damage minimised. Standards, therefore, specify maximum value of final setting time.

It is not possible, however, in practice to exactly locate the initial setting time and final setting time. The Indian standards have selected two arbitrary points which relate strength of cement to time from adding water.

'Initial setting time' is defined as the period elapsing between the time when the water is first added to the cement and the time at which the needle of 1 mm square section (Fig. 8.1) fails to pierce the test block to a depth of about 5 mm from the bottom of mould. 30 minutes is the minimum initial setting time specified by ISI for ordinary, rapid hardening cement and other cements and 60 min for low heat cement (see Table).

The 'final settimt time' is defined as the period elapsing between the time when the water is added to the cement and the time at which the needle of 1 mm square with 5 mm dia attachment (shown in the Fig. 8.1) makes an impression on the test block, while the attachment fails to make an impression on the test block. 600 minutes is the maximum time specified for final set for all the above mentioned Portland cement (see Table).

Apparatus (Fig. 8.1)

- Vicat's apparatus with mould and non-porous plate (glass or metal)
- (i) Needle (A) as shown in the Fig. 8.1 (ii) Needle (B)
- Balance (with weight box) capacity 1 kg
- Trowel of about 210 g weight
- Enamel trough
- Standard spatula
- Stop watch
- Thermometer centigrade (0° to 100°)
- Measuring cylinder - 500 ml.

Procedure

(a) Weigh about 400 gms of neat cement.
(b) Prepare a neat cement paste by adding 0.85 times the percentage of water required for standard consistency.
(c) Start the stop watch at the instant when water is added to the cement.
(d) Fill the Vicat's mould with the cement paste prepared with the mould resting on the non-porous plate. Gauging time should not be less than 3 minutes and not more than 5 minutes.
(e) Fill the mould completely and smooth off the surface of the paste making it level with the top of the mould to give a test block.
(f) Place the test block confined in the mould and resting on the non-porous plate under the rod, bearing the needle (A).
(g) Lower the needle gently till it comes in contact with the surface of the test block and quickly release, allowing it to penetrate into the test block and note penetration, after every two minutes.
(h) Repeat this procedure until the needle fails to piece the block for about 5 mm, measured from the bottom of the mould. Stop the stopwatch and note the time, which is the initial setting time.

Determination of final setting time

(a) Replace the needle (A) of the vicat apparatus by the needle with an annular attachment.
(b) Go on releasing the needle as described earlier till the needle makes an impression thereon, while the attachment fails to do so.
(c) The time that elapses between the moment the water is added to the cement and when the needle only makes an impression is considered as final setting time for the cement under test.

Precautions

(a) Needle must be cleaned each time before use.
(b) Shift the position of the mould after recording the penetration reading so that the penetration may not be at the same place.
(c) Check up the stop watch for accuracy.
(d) Clean appliances shall be used for gauging.
(e) Test block should be kept in 90% relative humidity and at $27^{o} \pm 2^{o}C$ and away from draught.

Observation table

(a) Samples prepared on	(i) date	:	....
	(ii) time	:	....................hrs. ....
(b) Temperature at testing	(i) cement	:	.... ^{o}C
	(ii) water	:	.... ^{o}C
	(iii) room	:	.... ^{o}C
(c) Quantity of cement (C)	g	=	....
(d) Water for standard consistency (P)		=	....
(e) Water to be added 0.85 P × C		=	....

Sample I

S.No.	Time at which water is first added (T_1) hr	min	Time at which penetration readings are taken hr	min
1				
2				
3				
4				
5				

(f) Time when needle fails to penetrate 5 mm to 7 mm from the bottom of the mould (T_2) =

(g) Time at which water is first added (T_1) =

(h) Initial setting time ($T_2 - T_1$) =

(i) Final setting time

Sample I

S.No.	Record of observation Penetration reading (T_3) hr	min
1		
2		
3		
4		
5		

(j) Time when the needle makes an impression but the attachment fails to make one (T_3) =

(k) Final setting time ($T_3 - T_1$) =

Questions for discussion

1. Differentiate between initial and final setting time of cement?
2. Explain the significance of performing this test?
3. Differentiate between 'setting and hardening'?
4. Justify the following statement : "Low heat cement has more initial setting time as compared to other 'Portland cements'?

EXPERIMENT 3

FOR A GIVEN SAMPLE OF CEMENT, VERIFY THE SOUNDNESS OF CEMENT

Statement of problem

Two samples of cements are given, find out :

(i) which cement is unsound, and
(ii) which cement is sound.

Theory

It is essential that cement concrete does not undergo large change in volume after it has hardened. This is ensured by limiting the quantities of free lime (CaO) and magnesia (MgO) in cement. During manufacture of cement, free lime is produced. Free-lime reacts with water and increases in volume considerably. Magnesia also has the same effect but its rate of reaction is slow. Larger percentage of free lime and magnesia, if present, therefore, tends to increase the volume of the hardened concrete, thus causing disintegration. The cement is, therefore, said to be unsound when the percentage of free lime and magnesia is more than that specified by Bureau of Indian Standard.

Unsoundness is measured with the help of the Le-Chatelier mould as explained in the procedure.

It must be specially mentioned here that in the event of the cements failing to comply with the above requirement, a further test should be made by the "Le-Chatelier Method" from another portion of the sample after aeration by spreading it out to a depth of 75 mm at a relative humidity of 50 to 80 per cent for a total period of 7 days. The expansion of the aerated sample is determined with the "Le-Chatelier Method" which shall not be more than 5 mm as given in BIS.

Apparatus (Fig. 8.2)

- Le-Chatelier apparatus
- Glass sheet (2 Nos.)
- Trowel
- Measuring cylinder
- Blow lamp or stove or water bath with electric heating arrangement
- Enamel tray
- Thermometer
- Stop watch
- Balance and weight box

Procedure

(a) Place the mould on a glass sheet and fill it with cement paste formed by gauging cement with 0.78 P. Care should be exercised to keep the edges of the mould gently together, while this operation is being performed.
(b) Cover the mould with another piece of glass sheet.
(c) Place a small weight on this, covering glass sheet and immediately submerge the whole assembly in water at a temperature of 27° to 32°C and keep there for 24 hours.

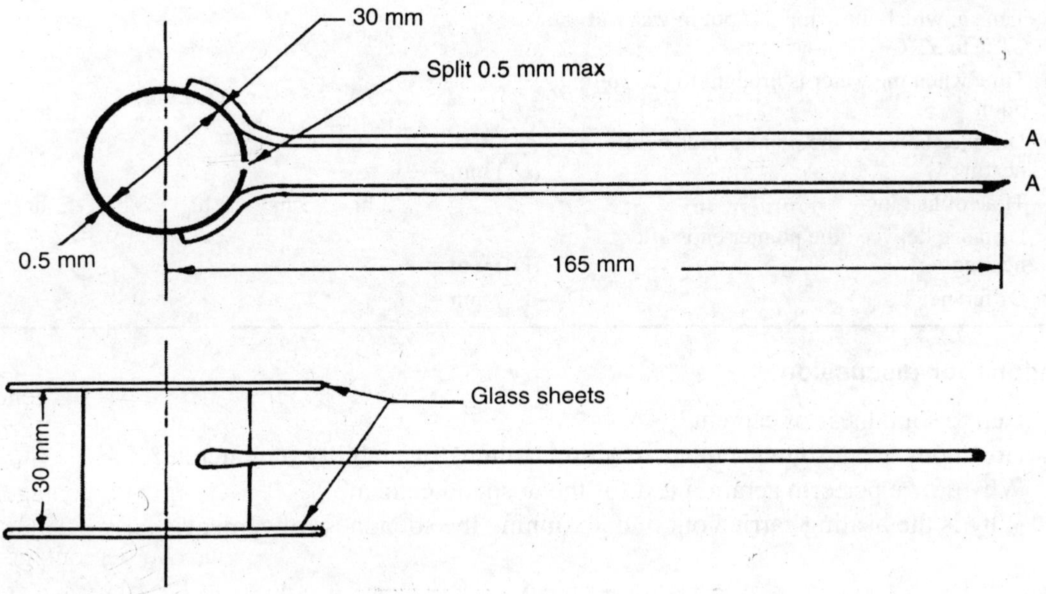

Fig. 8.2. Le-Chatelier's mould.

(d) Measure the distance separating the indicating points to an accuracy of a mm after 24 hours.
(e) Submerge the mould again in water at a temperature prescribed above.
(f) Bring the water to boiling, with the mould kept submerged in 25 to 30 minutes, and keep it boiling for three hours.
(g) Remove the mould from the water, allow it to cool and measure the distance between the indicator points.
(h) The difference between these two measurements represent the expansion of the cement.

Precautions

(a) All the measurements of quantity of cement and water should be done accurately by weight.
(b) The edges of the split mould should be kept together gently while filling the mould with the paste. Even a fine thread can be wound very gently so as to avoid the splitting up of the brass mould due to filling of the paste.
(c) Water should be brought to boiling point gradually in the specified time.
(d) Le-Chatelier apparatus should be handled by pressing indicator arms.

Observation table

		Before aeration			After aeration		
		I	II	III	I	II	III
(a) Weight of cement sample (C)	gm =						
(b) Water required for standard consistency	(P) =						
(c) Water added to the sample (0.78 P × C)	=						

Contd.

(d) Time at which the sample is put in water at 27°C to 32°C	=						
(e) Time when the water is brought to the boiling point	=						
(f) Distance between the pointer ends before heating	(D_1) mm =						
(g) Time of heating	=	3 hr	3 hr	3 hr	3 hr	3 hr	3 hr
(h) Distance between the pointer ends after heating	(D_2) mm =						
(i) Difference	($D_2 - D_1$) mm =						

Questions for discussion

1. Define soundness of cement?
2. How does the unsound cement affect the volumetric stability of concrete?
3. Why do we perform aeration test for the unsound cement?
4. Why is the heating carried out in determining the soundness of cement?

EXPERIMENT 4

DETERMINATION OF FINENESS OF CEMENT

Statement of problem

For a given sample of cement, determine the fineness of cement.

Theory

Strength development of concrete is the result of the reaction of water with cement particles. The reaction always starts with the cement available at the surface of the particles. Thus, larger the surface area available for reaction, greater is the rate of hydration. Rapid development of strength requires greater degree of fineness. Rapid Hardening Cement, therefore, requires greater degree of fineness.

However, too much fineness is also undesirable, because the cost of grinding the cement to higher fineness is considerable. Finer cement deteriorates more quickly when exposed to air and likely to cause more shrinkage, but less prone to bleeding. Greater fineness also requires greater amount of gypsum for proper retardation. Also amount of water requirement for the paste of standard consistency is greater.

It is, therefore, necessary to ensure certain amount of coarseness is the cement, but maximum limit to this coarseness is prescribed by BIS to ensure minimum degree of grinding.

Apparatus

- Weighing balance capacity 200 grams
- IS Test Sieve 90 microns

- Blaine's Air permeability apparatus
- Karandi or Trowel
- Tray 30 cm × 30 cm
- Bristle brush with 25 cm handle

Procedure

A. Fineness by sieving

(a) Weigh accurately 100 gm of cement and place it on a standard IS Sieve 90 microns.

(b) Breakdown any air set lumps in the sample with fingers, but do not rub on the sieve.

(c) Continuously sieve the sample by holding the sieve in both hands and giving a gentle wrist motion or mechanical sieve shaker may be used for this purpose. The sieving should be continuous for 15 minutes.

(d) Weigh the residue left after 15 minutes sieving. This residue shall not exceed the specified limits.

Precautions

(a) The cleaning of the sieve should be done very gently with the help of a brush i.e. 25 mm or 40 mm bristle brush with 25 cm handle.

(b) After sieving the cement must be removed from the bottom surface of the sieve gently.

(c) Weighing machine should be checked before use.

(d) Sieving must be carried out continuously.

Observation table

			Sample I	*Sample II*	*Sample III*
(a) Weight of cement	(W)	g	100	100	100
(b) I.S. sieve size		mic	90	90	90
(c) Sieving time		min	15	15	15
(d) Weight retained on sieve	(W_1)	g			
(e) Per cent weight retained on sieve	$\frac{W_1}{W} \times 100$	%			
(f) Mean percentage					

B. Fineness by air permeability method

Introduction

Blaine's air permeability apparatus is according to IS : 5516 for determining the fineness of portland cement in terms of the specific surface expressed as total surface area in square centimeters per gram of cement.

Description (Fig. 8.3)

The Blaine's air permeability apparatus consists of a permeability cell 1.25 cm ± 0.10 cm inside diameter. The top of the cell is at right angles to the principal axis of the cell. The bottom of the cell forms an airtight connection with the top of a manometer. A ledge forms an integral part of the cell at a depth of 5.0 cm ± 1.5 mm from the top for support of the metal perforated disc. The disc 0.9 cm ± 0.1 mm thick has 1 mm diameter perforations equally distributed. It fits inside the cell snugly.

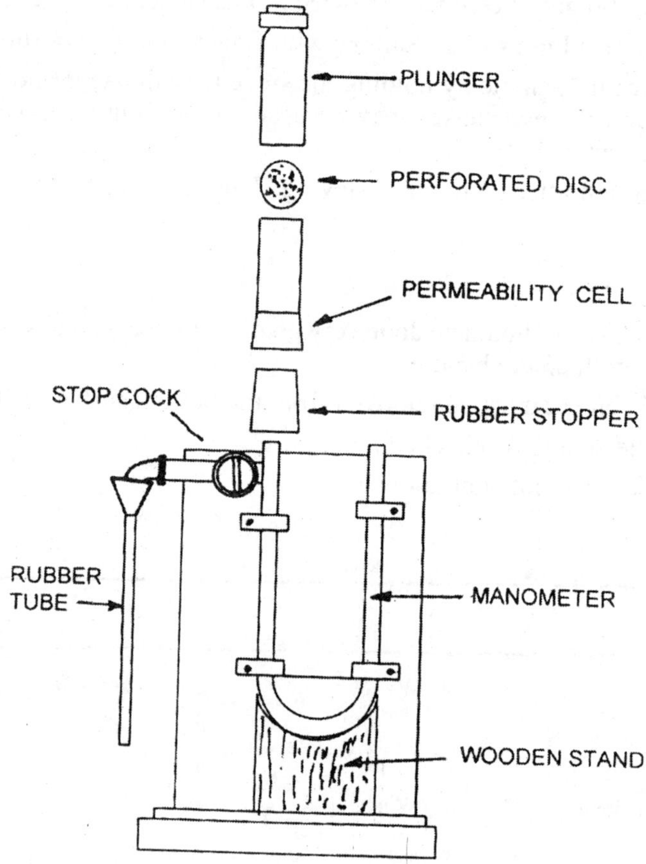

Fig. 8.3. Blaines air permeability apparatus.

The plunger which fits into the cell has sharp square edges. The top of the plunger is provided with a collar which when brought in contact with the top of the cell, the distance between the bottom of the plunger to the top of the perforated disc is 1.5 ± 0.1 cm.

The top of one arm of the U-tube manometer forms an airtight connection with the permeability cell. The manometer arm connected to the permeability cell has line markings around the tube below the top of the side outlet at a distance of 3.0 cm, 7.0 cm, 12.5 cm and 14.0 cm. A side outlet with a stop cock is provided near the top for use in depressing or raising the liquid in the manometer. The manometer is mounted firmly and in such a manner that the arms are vertical.

Setting up (refer Fig. 8.3)

Place the perforated disc inside the cell resting on the ledge. Insert the plunger into the cell. Fix up the rubber stopper on top of the manometer arm connecting the cell. Push in the stopper tight into the cell. Fill the manometer with dibutylphthalate.

Calibration

(a) Take a standard sample of cement (standard cement is one which has known specific surface area and specific gravity) at room temperature and determine the bulk volume of the compacted bed of powder by the mercury displacement method as follows.

(b) Place two filter paper discs in the permeability cell, and press it down with a rod slightly smaller than the diameter until the filter discs are flat on the perforated metal disc.

(c) Apply oil to the interior of the cell.

(d) Now fill it with mercury removing any air bubbles adhering to the wall of cell, level the mercury with the top of the cell by means of a small glass plate. Remove the mercury from the cell, weigh and record the weight of the mercury as 'W_A' gms.

(e) Remove one of the filter paper discs and put 2.8 gms of standard cement sample into the cell.

(f) Place the filter paper disc on top of the cement bed. Now insert the plunger and compress until the plunger collar is in contact with the top of the cell.

(g) Slowly remove the plunger and fill the space with mercury and level it with the glass plate. Now remove the mercury from the cell, weigh it and record its weigh as 'W_B' gms.

(h) Calculate the bulk volume occupied by the cement from the formula.

$$V = \frac{W_A - W_B}{D}$$

where:

V Bulk volume of cement in cubic centimeters

W_A Grams of mercury required to fill cell, no cement being in cell

W_B Grams of mercury required to fill the portion of the cell not occupied by the prepared bed of cement in cell

D Density of mercury at test temperature in grams per cubic centimeter (see Table 8.1).

Having calculated the bulk volume of cement in cubic centimeters proceed with the calibration.

(i) Take the contents of a vial of the standard cement sample into a jar and shake vigorously for 2 minutes to fluff the cement and break up any lumps. Calculate the weight of the standard cement sample, such that the porosity of the bed of cement is 0.500 ± 0.005.

$$W = \rho V (1 - e)$$

where

W weight of standard cement sample required

ρ specific gravity of test sample (for portland cement a value of 3.15 shall be used, for standard cement sample specific gravity value shall be provided by the manufacturer).

V bulk volume of bed of cement in cubic centimeters as determined

e desired porosity of bed of cement (0.500 ± 0.005).

(j) Keep the perforated disc on the ledge with a filter paper disc over it. Weigh the standard cement sample as calculated above to the nearest 0.001 gm and place it in the cell. Compress as described earlier with the plunger till the plunger is in contact with the top of the cell. Remove the plunger slowly.

(k) Moisten the manometer coupling with water before connecting the permeability cell. Attach the cell to the manometer tube making sure that an airtight connection has been made taking care that the prepared bed of cement doesn't get disturbed. Determine the efficiency of the connection by stoppering the top of the cell attached to the manometer and partially depressing the liquid in one arm of the manometer and then closing the stop cock. Any continuous drop in the pressure indicate a leak in the system.

(l) Now with the apparatus in experimental position (with no plunger on top) depress the liquid in the arm of the manometer upto the bottom most mark (4) and start the stop watch when the meniscus of the liquid reaches mark (3). Stop the stop watch when the liquid reaches the second mark and stop it as the bottom of the meniscus reaches this mark (2). Record the time interval in seconds. Repeat the calibration thrice with three different standard samples. Calibration shall be made by the same operator who makes the fineness test. The value of the constant 'K' can then be determined by the following relation :

$$K = \frac{S_s}{\sqrt{T_s}}$$

where :

S_s specific surface area of the standard cement sample (supplied by the manufacturer)

T_s time interval measured in seconds for the standard cement sample.

The value of constant 'K' so determined shall be same for the given apparatus and accessories.

Specimen preparation and test procedures

(a) Calculate the weight of the cement sample using the formula :
$w = \rho v\,(1 - e)$
where the value of 'ρ' shall be taken as 3.15, 'v' as calculated previously 'e' = 0.500 ± 0.005 (for all portland cement).

(b) For calculating the weight of cement sample for high early strength cement the weight of the sample shall be calculated taking 'e' as 0.530 ± 0.005.

(c) After having calculated the weight of given cement sample, weigh it and compress it with plunger as described earlier under calibration.

(d) Perform the experiment as described earlier and note down the time taken by the liquid to rise from the level (3) to level (2). Let this time be 'T' in secs. The specific surface area for the cement sample can thus be calculated as :

$$S = \frac{S_s}{\sqrt{T_s}} \times \sqrt{T}$$

or $\quad S = K.\sqrt{T}\ \text{cm}^2/\text{gm}$

Table 8.1. Density of mercury at different tempe-ratures

Room temperature degree centigrade	*Density of mercury g/cm^3*
16	13.56
18	13.55
20	13.55
22	13.54
24	13.54
26	13.53
28	13.53
30	13.52
32	13.52
24	13.51

Precautions

(a) All the connections should be air tight.
(b) The liquid in the U-tube manometer should be free from any impurities.
(c) The lower meniscus of the liquid should be taken into account while noting the time.
(d) The value of the constant 'K' is true for given set of apparatus and accessories.

Observation Table

1. Weight of Mercury in cell (full) 'W_A' = gms
2. Weight of Mercury in cell not occupied by prepared bed of cement 'W_B' = gms
3. Room temperature = °C
4. Density of Mercury =
5. Bulk volume of cement $V = \dfrac{W_A - W_B}{\text{Density}}$ = cm^3
6. Weight of sample required = ρv (1 – e) = gms
7. Weight of high early strength cement (using e = 0.530 ± 0.005) = ρv (1 – e) = gms
8. Specific surface area $S = \dfrac{S_s}{\sqrt{T_s}} \sqrt{T}$ = cm^2/gm

T_s - measured time interval for the standard sample
S_s - specific surface area for the standard sample

Questions for discussion

1. What do you mean by "fineness of cement"?
2. How is strength of cement related to fineness of cement?
3. How does air permeability test differ from fineness by sieving?
4. Study the fineness limits for different types of cement gives in part I of this manual?

EXPERIMENT 5

TO DETERMINE THE COMPRESSIVE STRENGTH OF CEMENT

Statement of problem

For the given sample of cement, determine the compressive strength of cement.

Theory

The compressive strength test is the final check on the quality of cement. The compressive strength is measured by determining the compressive strength of cement mortar cubes of 1 : 3 proportions. The fine aggregate used is the standard sand specified by IS 650. The compression test enables also to distinguish rapid hardening cement from low heat and ordinary cement. The O.P.C. is now being classified as 33 Gr. O.P.C., 43 Gr. O.P.C. and 53 Gr. O.P.C. depending on the strength of cement. C.R.R.I. has developed curves for concrete compressive strength v/s water cement ratio corresponding to the 7 and 28 days compressive strengths of cement. The compressive strength of cement enables us to distinguish cements of different strength and maximum use can be made of high strength cement.

Apparatus

- Cube vibration machine
- Cube moulds 7.06 cm
- Trowel
- Enamel trough
- Measuring cylinder 1000 cc
- Balance
- Thermometer
- Non-Porous plate

Procedure

(a) The material for each cube shall be mixed separately and quantities of cement, standard sand and water are as follows :

Cement 185 gm; Standard sand 555 gm.

Water P/4 + 3.0 per cent of combined weight of cement and sand.

(b) Place on a non-porous plate a mixture of cement and standard sand in the proportion of 1 : 3 by weight as given above.

(c) Mix it dry with a trowel for one minute and then add water until the mix is of uniform colour.

(d) Gauging time should not be less than 3 minutes and should not exceed 4 minutes. If it exceeds, mixture is rejected and operation is repeated.

(e) Oil the interior faces of the mould.

(f) Place the assembled mould on the table of the vibration machine and firmly hold it in position by means of suitable clamps.

(g) Securely attach a hopper of suitable size and shape at the top of the mould to facilitate filling and this hopper shall not be removed until completion of the vibration period.

(h) Immediately after mixing the mortar as specified above, place the entire quantity of the mortar in the hopper above the cube mould and compact the same by vibration for a period of about 2 minutes at a speed of 1200 $\pm$ 400 vibration/min.

(i) Keep the cubes at a temperature of $27^{o} \pm 2^{o}C$ in an atmosphere of at least 90% relative humidity for 24 hours after completion of vibration.

(j) At the end of this period remove them from the mould and immediately submerge in clean fresh water at $27^{o}C \pm 2^{o}C$ and keep there until taken out just prior to breaking. The water in which cubes are submerged should be renewed after every 7 days. The cubes should not be allowed to dry up before testing.

Testing

Test three cubes for compressive strength at the periods mentioned below, the period being reckoned from the completion of vibration.

Ordinary portland cement 3 and 7 days : rapid hardening portland cement 1 and 3 days; low heat portland cement 3, 7 and 28 days.

(i) Place the test cube on the platform of the compression testing machine without any packing between the cube and the steel plattens of the testing machine.

(ii) Apply the load steadily and uniformly starting from zero at a rate of 350 kg/cm^2/minute till the cube fails.

(iii) Calculate the compressive strength as specified under computations.

Precautions

(a) All appliances should be clean.

(b) The mixture which takes more than 4 minutes of mixing should be rejected.

(c) In assembling the mould, cover the joints between the halves of the mould with a thin film of petroleum jelly in order to ensure that no water escapes during vibration.

Observation table

(a) Sample prepared on — (i) date ....
(ii) time .... hrs. .... min.

(b) Temperature — (i) room .... ^{o}C
(ii) water .... ^{o}C

		Specimens					
		I	II	III	IV	V	VI
(c) Weight of cement	g						
(d) Weight of standard sand	g						
(e) Weight of water (P/4 + 3.0) × wt. of dry aggregate and cement							

(f) Size of sample (L) cm
(B) cm
(h) cm

(g) Area of specimen (A) cm^2

(h) Load at fracture (P)

Age :

1 day	kg
3 days	kg
7 days	kg
28 days	kg

(i) Compressive stress

Age :

1 day	kg/cm^2
3 days	kg/cm^2
7 days	kg/cm^2
28 days	kg/cm^2

(j) Average strength :

	1 day	3 days	7 days	28 days
kg/cm^2				

Conclusion

Referring to the specifications of cement, the test results indicate that the cement satisfies/does not satisfy the criteria.

Questions for discussion

1. Define compressive strength of cement?
2. Explain the effect of temperature on development of compressive strength of cement?
3. What effect fineness of cement has on compressive strength of cement?
4. Explain the effect of tricalcium silicates and dicalcium silicate on compressive strength development of cement?

Experiment 6

Field Test to Verify Quality of Cement

Statement of problem

Four samples of cements are given, find out :

(i) Which cement has setting and hardening action impaired.
(ii) Which cement is adulterated.
(iii) Which cement is unsound.
(iv) Which cement is of good quality.

Apparatus

- Stove
- Enamelled trays
- Test tube
- Glass tumbler
- Measuring steel rule

Procedure

A. Setting and hardening action

(a) Prepare three small pats, each 75 mm × 75 mm × 25 mm in size from the sample given with 28 per cent water by weight.
(b) Prepare similar number of pats with good quality cement.
(c) Cover the pats with moist cloth for 24 hours.
(d) Try to make thumb nail impression. Good quality cement will resist this impression.
(e) If the cement does not resist this impression, then continue curing up to 48 hours, after which try to break it with pressure of thumb. Bad quality cement will easily break under the pressure. Such cement should be tested in the laboratory.
(f) If 48 hour-test show improvement in hardening, but does not attain hardness comparable with genuine cement, further trial should be made after 72 hours of curing. If the only defect in the cement under test is its slow-setting quality, it will become as strong as the genuine cement in this third test.

B. Detection of adulteration

(a) Take a small sample of doubtful cement on a steel plate and heat it thoroughly to 20 minutes on a stove. Adulterated cement will change its colour on heating. In genuine variety there will be no change in colour.
(b) To detect adulteration with coal ash take a small quantity of doubtful cement in test-tube or a glass tumbler and add water till the glass container is half full.
(c) Shake the container thoroughly and allow it to settle for a few minutes.
(d) Cement particles will settle down and ash particles will either be found floating on the surface or held in suspension, because of their lightness than cement particles.

C. Ascertaining soundness of cement

(a) Make a pat of cement 75 mm in diameter and 15 mm thick and cure it with moist cloth for 24 hours and then boil in water for a period of 6 hours.
(b) Observe the surface of the pat. If the cement is sound, the surface will not develop a pattern of cracks as shown in Fig. 8.4(c). These cracks are thin and uniformly distributed all over the surface.

Precautions

(a) In a test for soundness of cement, the cracking of unsound cement should not be confused with contraction cracks.

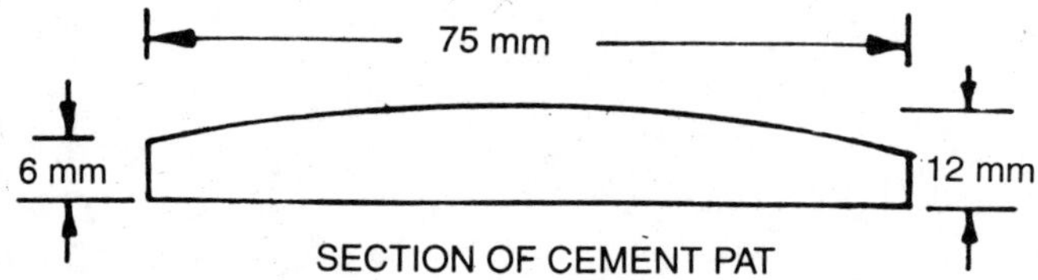

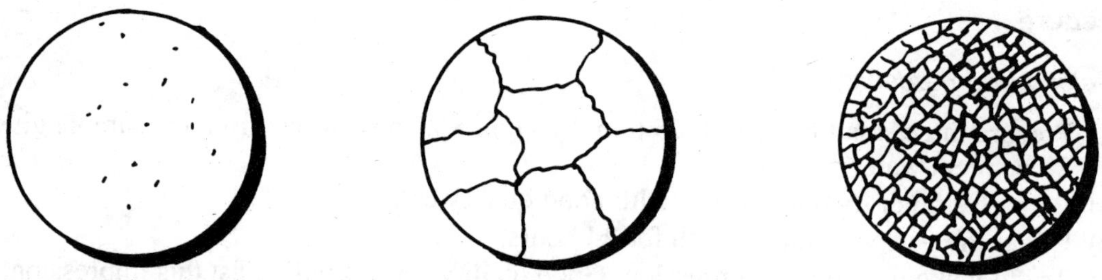

(a) No cracks appear on surface after boiling

(b) Surface between these cracks does not show crazing (minute crack)

(c) Crazing all over the surface due to expansion of unsound cement

Fig. 8.4. Field test for soundness of cement.

(b) Contraction cracks develop during boiling where the cakes might have been exposed to heat or drying winds. Contraction cracks are few well-defined cracks running from edge to edge as shown in Fig. 8.4(b) and they do not indicate anything wrong with the sample.

Note for teachers

While giving sample for test, give samples which are artificially adulterated or made unsound, so that the student should see the actual behaviour of adulterated cement.

Observation table

(a) Test results for setting and hardening.
(b) Test results for adulteration.
(c) Test results for soundness.

CHAPTER 9

Tests on Aggregate

EXPERIMENT 1

VISUAL INSPECTION OF AGGREGATE

Statement of problem

For two samples of aggregate, make a visual inspection of both.

Theory

Visual inspection is necessary to get an overall idea of aggregate received on site. This also helps the laboratory worker to get the idea of the aggregate received on site and plan the mix design and laboratory test procedures.

Procedure

(a) Obtain representative sample of about 2 kg.
(b) Locate the rock group to which it belongs. The broad rock-groups are mentioned in IS:388 and are reproduced partly in part I. It should normally be got done with the help of an expert geologist.
(c) Put a large drop of water on the aggregate and judge the absorption capacity of the aggregate by the speed with which the water is absorbed (state whether the absorption is nil, weak, strong, very strong etc.).
(d) Feel with thumb the texture and classify it.
(e) Observe the shape, size and cleanliness of aggregate.
(f) Final judgement should depend upon the visual inspection of large number of particles.

Note for teachers

In order to enable students to discriminate, teachers should preferably supply two samples viz.:

(i) 40 mm size graded dusty angular aggregate with medium absorption.
(ii) 20 mm single size rounded clean aggregate with no absorption.

Teacher may use other combinations also.

Observation table

(a) Work site	....
(b) Supplier and name of quarry	....
(c) Type of aggregate	....
(d) Date of supply	....
(e) Rock group	....
(f) Water absorption	Nil/Weak/Average/Strong
(g) Surface texture	Glossy/smooth/granular/crystalline/fine/medium/coarse/honey combed
(h) Shape	Rounded/irregular/angular/flaky or elongated
(i) Cleanliness	Clean/coated with thin layer of dust/coated with large amount of dust

Questions for discussion

1. In what way shape and size of aggregate affect workability of concrete mix?
2. What are the major rock groups to which coarse aggregates belong?
3. Does surface texture affect strength of concrete? If so, how?
4. Why is visual inspection of aggregates necessary?

Experiment 2

To Determine Flakiness Index and Elongation Index of Coarse Aggregate

Statement of problem

Given two samples of coarse aggregate of the same nominal size, determine which of the two is acceptable for concreting work.

Theory

The aggregates which are flaky or elongated are not desirable because the large number of flaky aggregates create larger percentage of voids, which inturn will require larger amount of fine materials and also more water for the same workability, since such shapes offer more surface area for lubrication. It also tends to harm durability, because the particles tend to be oriented in one plane thus causing laminations.

The aggregate is said to be flaky when its least dimension is less than 3/5th of its mean dimension. Mean dimension is the average of sieve size through which the particles pass and the sieve size on which it is retained. For example, a mean sieve size of the particle passing through 25 mm and retained on 20 mm is 20 + 25/2 = 22.5 mm. If the least dimension is less than 3/5 × 22.5 = 13.5 mm, then material is classified as flaky.

The material is said to be elongated when its length is greater than 9/5 times the mean sieve size.

Flakiness index or elongation index is the weight of flaky or elongated particles measured as percentage of the total weight of the sample. Flakiness index in excess of 35 to 40 per cent is considered undesirable. No limits for elongation index are known to have been prescribed as yet.

Apparatus

(i) Balance, (ii) Metal gauge, (iii) I.S. sieves of sizes as shown below :

Table 9.1. Dimension of thickness and length gauges

Size of aggregate passing through	*Retained on IS sieve*	** Thickness gauge mm*	** Length gauge mm*
63 mm	50 mm	33.90	—
50 mm	40 mm	27.00	81.00
40 mm	31.5 mm	19.50	58.5
31.5 mm	25 mm	16.95	—
25 mm	20 mm	13.50	40.5
20 mm	16 mm	10.9	32.4
16 mm	12.5 mm	8.55	25.6
12.5 mm	10 mm	6.75	20.2
10 mm	6.3 mm	4.89	14.7

Example :

$$\text{* Thickness gauge size } \frac{3}{5} \times \frac{(50+40)}{2} = 27.0 \text{ mm}$$

$$\text{* Length gauge size} = 1\frac{4}{5} \times \frac{(50+40)}{2} = 81.0 \text{ mm}$$

Thus the value of 27.0 mm for thickness gauge and 81.0 mm for length gauge has been prescribed for aggregate passing through 50 mm sieve and retained on 40 mm size.

Procedure

A. For flakiness index

(a) Take a quantity of aggregate sufficient to provide minimum number of 200 pieces of any fraction to be tested.

(b) The sample is sieved through I.S. sieves as specified in the Table 9.1, e.g. 20 mm size aggregate will have to be sieved through following sieves:
20 mm, 16 mm, 12.5 mm, 10 mm and 6.3 mm.

(c) Separate the aggregate retained on the sieves mentioned in the Table 9.1.

(d) Try to pass each aggregate through the corresponding slot in the thickness gauge e.g. the material passing through 20 mm and retained on 16 mm sieves should be made to pass only through 10.9 slot. If the material passes through the slot, it is considered flaky.

(e) Weigh all the pieces which pass through the slot, since they are considered flaky.
(f) Calculate flakiness index as follows :

$$\text{Flakiness index} = \frac{\text{Weight of material passing through thickness gauge}}{\text{Total weight of aggregate}} \times 100$$

B. For Elongation Index

(a) Sieve the aggregate through corresponding sieves.
(b) Separate the aggregate retained on sieves.
(c) Try to pass each aggregate piece through the corresponding gauge size. If the length of the particle is such that it does not pass through the length gauge, it is said to have been retained by the length gauge.
(d) Weigh all the material retained by the length gauge.
(e) Calculate elongation index as follows :

$$\text{Elongation index} = \frac{\text{Weight of material retained by the gauge}}{\text{Total weight of the aggregate}} \times 100$$

Note for teachers

At least one sample should be so prepared that the flakiness index should be more than 35%.

OBSERVATION TABLE

A. Weight of sample taken (200 pieces) W_1 = ... gm.

Size of aggregate		Corresponding thickness gauge size	Weight of aggregate passing thr. thickness		Corresponding length gauge size	Weight of aggregate retained on length gauge	
Passing through IS sieve	Retained on IS sieve						
1	2	3	4		5	6	
mm	mm	mm	gm		mm	gm	
			I	II		I	II
50	40	27			81.00		
40	31.5	19.5			58.5		
31.5	25	16.95			—		
25	20	13.50			40.5		
20	16	10.90			32.4		
16	12.5	8.55			25.6		
12.5	10	6.75			20.2		
10	6.3	4.89			14.7		
		Total of (4) W_2 = Flakiness index = $\frac{W_2}{W_1} \times 100$ =			Total of (6) W_3 = Elongation index = $\frac{W_3}{W_1} \times 100$ =		

Questions for discussion

1. What is the mean dimension for 20 mm and 16 mm size aggregate?
2. Define flakiness index and elongation index and what are the normal limits for both the indices?
3. When is aggregate said to have flaky shape and elongated shape?
4. Why are flaky and elongated materials treated as undesirable for concrete?
5. If we use such material, what can we do to improve it?

EXPERIMENT 3

FIELD METHOD TO DETERMINE FINE SILT IN AGGREGATE

Statement of problem

Given two samples of sand, determine which sample of sand is required to be washed.

Theory

There are three types of harmful substances normally found in aggregates. Firstly, there are impurities which interfere with the hydration of cement. Organic matter found in natural aggregates are products of decay of vegetable matter and is in the form of humus. All organic matter is not harmful and it is best to check its effects by making cube tests. In case of doubts, however, it is advisable to conduct a test described below to find out whether the organic matter is in sufficient quantity as to warrant further tests. Even if the test indicates that the organic matter is in large quantities, the aggregate is rejected only when the cube test with the aggregate in question shows lesser strength as compared with the cube strength of the same mix proportions but made with known quality of aggregate.

The second category consists of coatings which prevent development of good bond between aggregate and cement paste. This category comprises of clay and slit. Clay is a material of size below 0.002 mm and silt size varies between 0.002 mm to 0.06 mm. These materials may be in the form of coating or in loose form. In both the cases, the presence of large percentage of fine material is harmful because of there fineness, resulting in the larger surface area. Silt and fine dust increases the amount of water for given workability. There is a field method of determine volumetrically the percentage of fine material. If the percentage exceeds 6% by volume, then more accurate test (sedimentation method) as explained in IS : 2386 (Part II) should be conducted. The limits prescribed by IS are applicable to this sedimentation test.

Normally if the field test indicates percentage of silt exceeding 6% by volume then it is advisable to wash the sand.

All the tests for the impurities as mentioned in IS : 2386 (Part II) have not been reproduced in this manual. Only procedure for field method of determining silt content has been mentioned below.

Apparatus

- 250 ml measuring cylinder.

Procedure

(a) Prepare 50 ml of approximately 1% solution of common salt in water in a measuring jar.
(b) Add sand till the level reaches 100 ml mark.
(c) More solution is then added until the total volume of mixture in the cylinder is 150 ml.
(d) Cover the cylinder with the palm of the hand and vigorously shake the contents turning it upside down.
(e) Allow the contents to settle down for 3 hours.
(f) The silt will now settle in a layer above the sand.
(g) Percentage of silt is calculated.

			Samples		
			I	II	III
(a) Volume of sample	(V_1)	ml			
(b) Volume of silt after three hours	(V_2)	ml			
(c) Percentage silt by volume	$\left(\frac{V_2}{V_1} \times 100\right)$				
(d) Average value, per cent					

Conclusions

Since the percentage of silt exceeds/does not exceed 6 per cent, fine aggregate is required/not required to be washed.

Questions for discussion

1. Why is excessive presence of silt considered harmful to the concrete?
2. If the above experiment indicates about 12% silt content, would you consider the fine aggregate suitable for use?
3. What corrective measures would you suggest if silt content is too high?

EXPERIMENT 4

DETERMINATION OF SPECIFIC GRAVITY AND WATER ABSORPTION OF AGGREGATES

Statement of problem

Determine specific gravity and water adsorption value of a given sample of aggregate and compare it with the known value of aggregates.

Theory

IS : 2386 - Part III mentions two types of specific gravities viz. specific gravity and apparent specific gravity.

Specific gravity is the ratio of the weight of aggregate maintained for 24 ± ½ hours at temperature 100°C to 110°C to the weight of equal volume of water displaced by saturated surface dry aggregate. In other words, it means volume of water displaced is equal to the volume of solid matter of the aggregate and volume of all pores both impermeable and permeable.

Apparent specific gravity is the ratio of the weight of aggregate dried and maintained for 24 hours ± ½ hour at temperature 100°C to 110°C to the weight of equal volume of water displaced by the dry aggregate. The volume of water displaced is, therefore, equal to the water displaced by solid matter and only impermeable pores.

It is the specific gravity calculated on the saturated surface dry basis that is normally used in the calculation of absolute volumes of concrete and yield of concrete.

Majority of aggregates have the values between 2.6 to 2.9. Values lower than 2.6 however, do not necessarily mean that the aggregate is of lower quality. No minimum limit has been specified for specific gravity. In certain constructions like gravity dams where the weight of the concrete is important from the point of view of stability, minimum limit is specified.

Specific gravity is useful for calculating void content in aggregate. If the aggregate is dry, it absorbs water which does not take part in the reactions and also in lubrication of particles. Thus workability is likely to be reduced and w/c ratio is also altered. It is, therefore, always necessary to make allowance for water absorption while calculating total water to be added to the mixes.

The absorption value is the difference in the weight between the saturated surface dry aggregate and very dry sample expressed as percentage of dry weight of aggregate.

Specific gravity of aggregate can be used the same way as bulk density for judging quality by comparison with normal specific gravity for that type of aggregate.

Apparatus

- Balance : A balance or scale of capacity not less than 3 kg readable and accurate to 0.5 g and of such a type as to permit the weighing of the vessel containing aggregate and water.
- Oven : Well ventilated and fitted with control to maintain a temperature of 100°C to 110°C.
- Wide mouthed glass vessel such as jar of about 1.5 litre capacity, with a flat ground lip and a plane ground disc of plate glass to cover it giving a virtually water tight fit.
- Cloth : Dry soft absorbent clothes, each not less than 72 cm × 45 cm.
- A tray : Shallow tray of area not less than 325 cm^2.
- Container : An air tight container, large enough to take the sample.
- Wire basket of not more than 6.3 mm mesh or a perforated container of convenient size, preferably chromium plated and polished with wire hangers not thicker than one mm for suspending it from the balance.

Procedure

(a) Take two kg of sample.

(b) The sample is screened on a 10 mm IS sieve, thoroughly washed to remove fine particles of dust, drained, then placed in the wire basket immersed in distilled water and allowed to remain immersed at a temperature of 22° to 32°C for 24 ± ½ hours.

(c) Soon after immersion and again at the end of the soaking period air entrapped in or bubbles on the surface of the aggregate should be removed by gentle agitation. This is achieved by rapid clock-wise and anti-clockwise rotation of vessel between the operator's hands.

(d) The vessel is over filled by adding distilled water and the plane ground glass disc slid over the mouth so as to ensure that no air is trapped in the vessel. The vessel shall be dried on the outside and weighed (A).

(e) The vessel is emptied and the aggregate is allowed to drain out. Refill the vessel with distilled water. Slide the glass disc in position as before. The vessel is dried and weighed (B).

(f) The aggregate is placed on a dry cloth and gently surface dried with the cloth, transferring it to a second dry cloth when the first will remove no further moisture.

(g) It is then spread out not more than one stone deep on the second cloth, and left exposed to atmosphere away from direct sunlight or any other source of heat for not less than 10 minutes or until it appears to be completely surface dry (which with some aggregates may take an hour or more). The aggregate is turned over at least once during this period and a gentle current of unheated air may be used after first 10 minutes to accelerate the drying of the difficult aggregate.

(h) The aggregate shall be weighed (C).

(i) The aggregate is placed in the oven in the shallow tray, at a temperature of 100°C to 110°C for 24 hours ± ½ hours. It is then cooled in air tight container and weighed (D).

(j) Determine specific gravity and water absorption value.

			Sample I	
			I	II
(a) Weight of sample taken		g		
(b) Weight of saturated and surface dry aggregate	(C)	g		
(c) Weight of pycnometer + sample + water	(A)	g		
(d) Weight of the pycnometer + water	(B)	g		
(e) Weight of oven dry sample	(D)	g		
(f) Specific gravity $\dfrac{D}{C-(A-B)}$				
(g) Apparent specific gravity $\dfrac{D}{D-(A-B)}$				
(h) Water absorption percentage dry weight $\dfrac{C-D}{D} \times 100\%$				
(i) Average values : Specific gravity Apparent specific gravity Water absorption				

EXPERIMENT 5

TO DETERMINE THE SPECIFIC GRAVITY AND WATER ABSORPTION WHEN THE SIZE OF THE AGGREGATE IS SMALLER THAN 10 MM (PYCNOMETER METHOD)

Statement of problem

Determine specific gravity and water absorption value of a given sample of aggregate and compare it with the known value of aggregates.

Theory

Same as described in Experiment No. 4.

Apparatus

- Balance : Same as used for the aggregate between 40 mm to 10 mm.
- Oven
- Pycnometer (Fig. 9.1)
- Air dried
- A tray of area not less than 325 cm^2
- An air tight container larger enough to take up the sample
- Filter paper and funnel

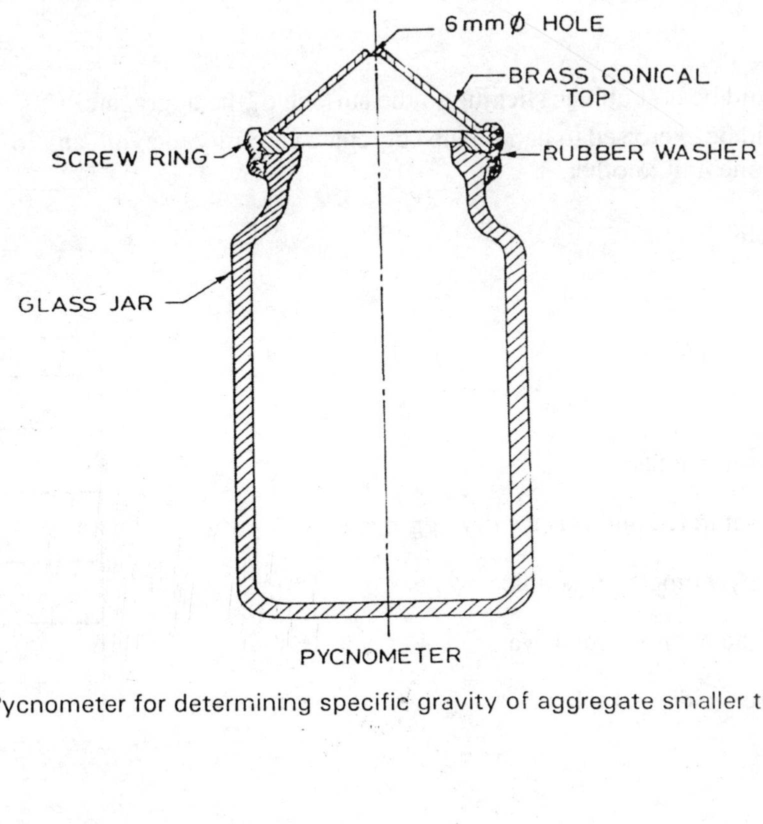

Fig. 9.1. Pycnometer for determining specific gravity of aggregate smaller than 10 mm.

Procedure

(a) A sample or about 1 kg for 10 mm to 4.75 mm or 500 g if finer than 4.75 mm is placed in the tray and covered with distilled water at a temperature of 22°C to 32°C.

(b) Soon after immersion air entrapped in or bubbles on the surface of the aggregate are removed by gentle agitation with a rod. The sample shall remain immersed for 24 ± ½ hour.

(c) The water is then carefully drained off from the sample by decantation through a filter paper and any material retained being returned to the sample.

(d) The aggregate including any solid matter retained on the filter paper is exposed to a gentle current of warm air to evaporate surface moisture and is stirred at frequent intervals to ensure uniform drying until no free surface moisture can be seen.

(e) The saturated and surface dry sample is weighed (Weight A).

(f) The aggregate is then placed in pycnometer (see Fig. 9.1) and then it should be filled with distilled water.

(g) Any air entrapped is eliminated by rotating the pycnometer on its side, the hole of the apex of the cone covered with finger.

(h) The pycnometer is then dried on the outside and weighed (B). The contents of the pycnometer are then transferred into a tray, care being taken to ensure that all the aggregate is transferred.

(i) The pycnometer is then refilled with distilled water to the same level as before, dried on the outside and weighed (C).

(j) The water from the sample is removed by decantation and the sample is dried in the oven at 100°C to 111°C for 24 ± ½ hours, cooled and weighed (D).

Precautions

(a) There should be no bubbles sticking to the surface of the aggregate.

(b) Care should be exercised to handle the aggregate carefully to avoid any loss during transferring from one unit to other.

Observation table

			Sample I	
			I	II
(a) Weight of sample taken		g		
(b) Weight of saturated and surface dry aggregate	(C)	g		
(c) Weight of pycnometer + sample + water	(A)	g		
(d) Weight of the pycnometer + water	(B)	g		
(e) Weight of oven dry sample	(D)	g		

(f) Specific gravity $\dfrac{D}{C-(A-B)}$		
(g) Apparent specific gravity $\dfrac{D}{D-(A-B)}$		
(h) Water absorption percentage dry weight $\dfrac{C-D}{D} \times 100\%$		
(i) Average values : Specific gravity Apparent specific gravity Water absorption		

EXPERIMENT 6

DETERMINATION OF BULK DENSITY AND VOIDS OF AGGREGATES

Statement of problem

Given following types of sample :

(a) 20 mm graded and angular aggregate.

(b) 20 mm single size and angular aggregate.

(c) 20 mm graded and rounded aggregate.

(d) Fine aggregate.

Determine the bulk density of each and explain why bulk-density in each case is different from the other. For a concrete proportion of 1 : 2½ : 5 by weight, calculate the quantity of coarse aggregate and fine aggregate by volume per bag of cement, if 20 mm (graded and angular) size coarse aggregate and fine aggregate given in the experiment are used.

Theory

Bulk density is the weight of aggregate required to fill a container of unit volume. This unit volume, therefore, consists of volume of solid material plus the volume of voids and is measured in kg/litre. These values are required to convert the quantity of aggregate by weight to quantities by volume when volume batching is done and vice versa.

Value of the bulk density of the aggregate depends upon the amount of effort used to fill the container as densely as possible, size distribution, shape and specific gravity. More graded the aggregate, greater is the bulk density. Angular and flaky shape of the material reduce the bulk density.

If this bulk density test is carried on frequently on the site, the appreciable change in the value of the bulk density at any one time helps to detect the change in grading or the shape of the material and enables the engineer on site to conduct further elaborate test, if necessary.

For batching purposes where the materials are measured, the bulk density of the 'loose' material should be calculated.

When the bulk density test is carried out to detect the change in grading and shape, the rodded bulk density test will have to be done to compare the results. Also for comparison of results size of two aggregates to be compared should be the same.

Apparatus

- Balance sensitive up to 0.5% of the weight to be weighed.
- Cylindrical metal measure 3, 15 and 30 litres capacity according to maximum size of the coarse aggregate.

Coarse aggregate	Measure
4.75 mm and less	3 litres
4.75 mm to 40 mm	15 litres
above 40 mm	30 litres

- Tamping rod 16 mm dia and 600 mm long with one end rounded.
- Container, trough, steel rule and measuring cylinder 250 ml.

Procedure

A. Rodded weight

(a) *Condition of the specimen* : The test is normally carried out on dry material when determining the voids, but when bulking tests are required material with a given percentage of moisture may be used. The mention of the condition must be made while noting observations.
(b) The aggregate is now tamped with 25 strokes of the rounded end of the tamping rod.
(c) A further similar quantity of aggregate is added and a further tamping of 25 strokes given.
(d) The measure is now finally filled to over-flowing, tamped 25 times and the surplus aggregate struck off, using the tamping rod as straight edge.
(e) The net weight of the aggregate in the measure is determined and the bulk density is calculated in kg/litres.

B. Loose weight

(a) The measure is filled to over-flowing by means of a shovel or scoop, the aggregate being discharged from height not exceeding 5 cm above the top of the measure. Care should be taken to avoid segregation of the particles.
(b) The surface of the aggregate is then levelled with a straight edge.
(c) The net weight of the aggregate in the measure is determined and the bulk density calculated in kg/litre.

Observation table

Condition of aggregate (air dry/surface dry/moist)

			Sample I	
			I	II
(a) Capacity of measure	(V)	l		
(b) Weight of measure	(W_1)	kg		
(c) Weight of measure + compacted aggregate	(W_2)	kg		
(d) Weight of measure + loose aggregate	(W_3)	kg		
(e) Rodded bulk density $\left(\dfrac{W_2 - W_1}{V}\right)$		kg/l		
(f) Average	(γ_r)	kg/l		
(g) Loose bulk density $\left(\dfrac{W_3 - W_1}{V}\right)$		kg/l		
(h) Average	(γ_r)	kg/l		
(i) Specific gravity (Gs)				
(j) Voids, $\dfrac{G_s - \gamma}{Gs} \times \dfrac{100}{1}$ per cent				

Questions for discussion

1. Define bulk density?
2. What are the factors that affect the value of bulk density?
3. What is the difference between 'loose bulk density' and 'rodded bulk density'?
4. During the test cn site, if there is appreciable change in the value of rodded bulk density, what will you conclude from it?

EXPERIMENT 7

DETERMINATION OF SURFACE MOISTURE IN FINE AGGREGATE BY 'DISPLACEMENT METHOD'

Statement of problem

Determine the surface moisture in a given sample of fine aggregate.

If the proportion by volume of concrete is 1 : 2½ : 5 with w/c = 0.7, calculate how much water per bag of cement should be added at the mixer, if the sand contained moisture as obtained in the experiment.

Theory

Aggregate is said to be saturated and surface dry, if all the pores in the aggregate are filled with water, but its surface is dry and contains no free moisture. This is the condition when the aggregate neither absorbs water nor does it give out moisture.

Surface moisture is expressed in terms of weight of saturated surface dry aggregate as well as in terms of weight of wet aggregate.

Aggregate, especially fine aggregate when exposed to rain collects large amount of moisture on the surface and it keeps this moisture over long periods. Fine aggregate received fresh from river bed also contains appreciable amount of moisture. Sand is sometimes required to be washed for removal of silt and then used. Such sand also contain moisture.

This moisture always adds to the water added in the mixer thus increasing water/cement ratio and thereby reducing the strength. Actual water to be added in the mixer should be the water calculated from the prescribed water/cement ratio less the free moisture content of the aggregate.

Frequently of conducting tests has been explained in the chapter 'Control tests for quality concrete'.

Apparatus

- Balance with capacity 2 kg or more and sensitive to 0.5 kg or less.
- Flask of glass suitably graduated. The volume of the flask should be from 2 to 3 times the loose volume of the sample. The container should be so designed that it can be filled up the mark, or the volume of its contents read within 0.5 ml, or less.

Procedure

(a) A representative sample of the fine aggregate to be tested for surface moisture contents should be selected and weighed (M_s). It should weigh not less than 200 g. Larger samples will yield more accurate results.

(b) Fill the container upto the mark with water and the weight in grams is determined (M_c).

(c) Empty the container and then add some water sufficient to cover the sample.

(d) Sample is now introduced into the container and any entrained air is removed.

(e) Fill the container with water up to the original mark and the weight in grams is determined (M). The amount of water displaced by the sample is calculated as follows :

$$V_s = M_c + M_s - M$$

where :

V_s is the weight in grams of water displaced by the sample.

M_c is the weight in grams of container filled up to the mark with water.

M_s is the weight of the sample in grams.

M is the weight in grams of the sample and container filled to the mark with water.

Calculations

$$P_1 = \frac{V_s - V_d}{M_s - V_s} \times 100$$

$$P_2 = \frac{V_s - V_d}{M_s - V_d} \times 100$$

where

P_1 – Percentage surface moisture in terms of saturated surface dry fine aggregate.

V_d – The weight of the sample divided by the specific gravity on saturated and surface dry basis.

P_2 – Percentage surface moisture in terms of the weight of wet fine aggregate.

Precaution

The accuracy of result depends upon the accurate information on the specific gravity of the material in saturated surface dry condition.

Observation table

(Displacement method)

			Sample I	
			I	II
(a) Weight of sample	(M_s)	g		
(b) Weight of container filled upto the mark with water	(M_c)	g		
(c) Weight of sample and container filled to the mark with water	(M)	g		
(d) Weight of water displaced $V_s = M_c + M_s - M$		g		
(e) $V_d = \frac{M_s}{\text{specific gravity of the sample}}$				
(f) Percentage surface moisture in terms of saturated and surface dry aggregate weight $P_1 = \frac{V_s - V_d}{M_s - V_s} \times 100$ per cent				
(g) Percentage surface moisture in terms of weight of wet aggregate $P_2 = \frac{V_s - V_d}{M_s - V_d} \times 100$ per cent				
(h) Average P_1				
P_2				

Questions for discussion

(See under Experiment No. 8)

EXPERIMENT 8

DETERMINE SURFACE MOISTURE IN FINE AGGREGATE (FRYING PAN METHOD)

Statement of problem

Determine the surface moisture in a given sample of fine aggregate, using frying pan method.

Theory

Same as described in Displacement methods (Experiment No. 7).

Apparatus

- Tray 60 cm × 30 cm
- Balance of 2 kg capacity
- Heater thermostatically controlled

Procedure

(a) Take about one kg of fine aggregate and weigh it (W_1).
(b) Dry the aggregate on a tray over a source of heat.
(c) Take care not to over-dry the aggregate. Heat it till it is brought to the free flowing condition. This stage is determined by feel or by forming the sand into a pile by means of a conical mould. When the mould is removed the material should slump freely. If the sand has acquired a brownish tinge, it is sure sign that excessive drying has taken place.
(d) Weigh the dried aggregate (W_2).
(e) Loss of weight is then calculated and expressed in terms of percentage of dried aggregate or wet aggregate as follows :

$$P_1 = \frac{W_1 - W_2}{W_2} \times 100 \text{ (in terms of dry aggregate)}$$

$$P_2 = \frac{W_1 - W_2}{W_2} \times 100 \text{ (in terms of dry aggregate)}$$

Observation table

(Frying Pan Method)

			Sample I	
			I	II
(a) Weight of wet sample	(W_1)	g		
(b) Weight of dry sample	(W_2)	g		
(c) Loss of weight ($W_1 - W_2$)		g		
(d) Percentage moisture in terms of dry aggregate $P_1 = \frac{W_1 - W_2}{W_2} \times 100$		%		
(e) Percentage moisture in terms of wet aggregate $P_2 = \frac{W_1 - W_2}{W_1} \times 100$		%		
(h) Average P_1				
P_2				

Questions for discussion (for Experiment No. 7 and 8)

1. Define 'Saturated and Surface dry' condition of aggregate?
2. How is surface moisture expressed?
3. How does surface moisture affect water/cement ratio?
4. In the field, when will you consider it necessary to conduct field-test for determining surface moisture?

EXPERIMENT 9

TO DETERMINE NECESSARY ADJUSTMENT FOR BULKING OF FINE AGGREGATE (FIELD METHOD)

Statement of problem

Given two samples of fine aggregate belonging to Zone I and Zone IV with same moisture content in each :

(i) Determine the bulkage of each.

(ii) If the fine aggregates are to be used in 1 : 2 : 4 (by volume) concrete, what volume of each, per bag of cement will be required to be added?

Theory

Bulking is the increase in volume of a given volume of sand caused by films of surface water pushing the sand particles apart due to surface tension.

Find sand bulk much more than the coarse sand. With the moisture content of 5 to 6 per cent, the increase in the volume of sand may vary from 20 to 40 per cent, depending upon the fineness of sand. Normally beyond this percentage of moisture content, bulking tends to decrease with further addition of water. When the sand is fully saturated, the volume is approximately the same when measured dry and loose.

When the sand is batched by volume, bulking results in smaller weight of sand occupying the fixed volume measuring box. The concrete mix then becomes deficient in sand and the concrete may be prone to segregation. The yield of concrete also is reduced. Therefore, while adding fine aggregate to the concrete mix by volume, the specified volume of dry material is proportionately increased if the sand is moist.

Part I gives the table giving a rough guide to the bulking which coarse, medium and fine sands normally undergo.

Apparatus

The apparatus consists of the following :

- A container
- Steel rule
- Steel rod 6 mm diameter
- 250 cc measuring cylinder

Procedure

The procedure to be adopted may be varied but two methods are suggested below. Both depend on the fact that the volume of inundated sand is same when the sand is dry.

Method I

(a) Put sufficient quantity of sand loosely into a container until it is about 2/3rd full.
(b) Level off the top of sand pushing a steel rule vertically down through sand at middle to bottom. Measure height. Suppose this is (h).
(c) Empty the sand out of the container into another container.
(d) Fill the container with water to ¼th volume occupied by sand.
(e) Put back about half the sand and rod it with a steel rod 5 mm dia so that its volume reduces to minimum.
(f) Add remainder of sand and rod it in the same way.
(g) Smooth and level the top surface of the inundated sand and measure its depth at middle with steel rule. Suppose it is (h').
(h) Calculate the bulkage of sand which is equal to $\frac{h-h'}{h'} \times 100$.

Method II

(a) In a 250 ml measuring cylinder, pour damp sand consolidated by shaking until it reaches 200 ml mark.
(b) Take the sand out, first fill it with some water and refill it with sand and stir the sand well.
(c) It will be seen that sand surface is now below original level. Suppose surface is at mark Y ml.
(d) Calculate bulkage = $\frac{200-Y}{Y} \times 100$

Precautions

No sand should be lost in transferring from one container to another.

Observation table

Method I

			Sample		
			I	II	III
(a) Height of loose sand	(h)	mm			
(b) Height of saturated sand	(h')	mm			
(c) Percentage bulking $\frac{h-h'}{h'} \times 100$		%			
(d) Average value		%			

Method II

			Sample		
			I	II	III
(a) Volume of loose sand	(h)	mm	200	200	200
(b) Volume of saturated sand	(Y)	mm			
(c) Percentage bulking $\left(\frac{200 - Y}{Y}\right) \times 100$		%			
(d) Average value		%			

Conclusions

For a given mix of.. by volume, the actual proportioning should be... .

Questions for discussion

1. Under what circumstances does the sand bulk?
2. What happens to concrete if allowance for bulkage for moist sand is not made during volume batching? How will it affect mix proportions?
3. Is bulkage allowance necessary, if weigh-batching is resorted to?
4. How is bulking of sand expressed?

EXPERIMENT 10

DETERMINATION OF PARTICLE SIZE DISTRIBUTION (GRADING OF AGGREGATE) OF FINE, COARSE AND ALL-IN-AGGREGATES BY SIEVE ANALYSIS

Statement of Problem

A. Sieve analysis of coarse aggregate

Samples of following categories are supplied. Conduct sieve analysis and identify each category :

(a) 40 mm graded
(b) 20 mm graded
(c) 20 mm single size

B. Sieve analysis of fine aggregate

Identify to which zone, two given samples of fine aggregate belong.

C. Sieve analysis of all-in aggregate

Following three samples of all-in-aggregate are supplied. Conduct sieve analysis and identify each:

(a) 20 mm continuously graded.

(b) 20 mm graded, but certain sizes are missing.

(c) 20 mm graded, but certain sizes are in excess.

Theory

Aggregates can be classified into (a) single size aggregate, (b) graded aggregate and (c) all in aggregate. In single size aggregate, the size of particles predominantly belongs to one or two consecutive sieve-sizes only. Graded aggregate contains all sizes of particles in suitable proportions (see Table 2.6, Part I). All-in-aggregate is the combination of both coarse aggregate and fine aggregate, each type having again all sizes of particles in suitable proportions (see Table 2.7, Part I).

Coarse aggregate is the aggregate retained on 4.75 mm sieve. The aggregate passing through 4.75 mm is classified as fine aggregate.

Aggregate is said to be graded when it contains different sizes of particles in suitable proportions. The main advantage of graded aggregate is that it provides minimum voids; smaller particles filling in the voids created by larger particles. It also improves considerably the workability. Single size aggregate or the aggregate having one sized particle considerably larger than the other sizes makes the concrete harsh and does not, when worked with trowel, give a good surface finish.

Sieve analysis enables us to determine the proportion of different sizes of particles. Results of sieve analysis are given in terms of percentage of total aggregate passing through each size of sieve. To have a visual grasp of the grading, these results can be plotted on a graph, whose ordinate indicates percentage passing and abscissa indicates sieve sizes on logarithmic scale. There is no ideal grading as such. But there are limits placed on the grading, upper limit for finer grading and the lower limit for coarser grading. Finer the grading, greater is the water requirement resulting in the poor quality concrete. Coarser the grading, greater is the tendency to segregate. The most suitable grading is that which gives minimum number of fines sufficient to give the mix necessary cohesiveness.

The limits for 20 mm and 40 mm size coarse aggregate and fine aggregate have also been mentioned in Part I and reproduced graphically (Figs. 2.3(a) and 2.3(b)). Limits for all-in-aggregate have also been mentioned. Zones into which sands can be classified have also been given; greater the zone number, finer is the grading of the aggregate.

To obtain good concrete, effort must be made to get the combined grading of fine and coarse aggregate whose graph should lie in 'Good Zone' (Fig. 2.4 indicates good zone) which gives coarser grading with sufficient amount of fine particles for good workability. This is possible with fine aggregate belonging to Zone I and II. If the natural sand available lies in Zone III and Zone IV then probably combined grading of fine and coarse aggregate in 'tolerance zone' (Fig., indicates tolerable zone) will have to be chosen. Both these ranges have been indicated in Fig. in Part I.

Method of combining fine aggregate and coarse aggregate to give a mixture approximately to the one desired has been explained in experiment No. 11.

Apparatus

Sieves type	Sieve designation
• Square hole perforated plate, 30 cm diameter	80 mm, 63 mm, 40 mm, 20 mm, 16 mm, 12.5 mm and 10 mm
• Fine mesh, wire cloth, 20 cm diameter	10 mm, 4.75 mm, 2.36 mm, 1.18 mm, 600 micron, 300 micron, 150 micron, 75 micron

- Balance (to weigh 10 kg)
  Brushes with fine camel hair

Test procedure for coarse, fine and all in aggregate

(a) The sample is brought to an air-dry condition before weighing and sieving either by drying at room temperature or by heating at a temperature of 100°C to 110°C.

(b) The dried sample is weighed.

(c) The weighed sample is placed on the sieve and sieved successively on the appropriate sieves starting with the largest.

(d) Each sieve is shaken separately over a clean tray until not more than a trace passes, but in any case for a period of not less than 2 minutes. The shaking is done with a varied motion, backward and anti-clockwise. Lumps of fine material if present may be broken by gentle pressure with fingers against the side of the sieve.

(e) At the end of the sieving, 150 micron and 75 micron sieves are cleaned from the bottom by light brushing with fine camel hair brush.

(f) On completion of sieving, the material retained on each sieve together with any material cleaned from the mesh, is weighed.

Precautions

(a) Aggregate sample should be air-dried, to prevent clogging of sieves.

(b) Finer sieves should be cleaned gently with soft brushes.

(c) Care should be taken to see that the sieves are not surcharged. For coarse aggregate being sieved through 30 cm sieve size, weight of sample should not exceed 1 kg and for fine aggregate to the sieved with 20 cm sieve size weight of sample should not exceed 500 gm.

(d) Sieving should be done by giving varied motion so that each particle gets sufficient chance of passing through the sieve opening.

Observation table

(a) Sieve analysis results for fine aggregate

IS sieve size	Weight of aggregate retained			Total weight				gm
	I	II	III	Sum	Average sum	Percent retained	Cumulative % retained	Percent passing
1	2	3	4	5	6	7	8	9
10 mm								
4.75 mm								
2.36 mm								
1.18 mm								
600 mic								
300 mic								
150 mic								
75 mic								
Pan								

Conclusion

(i) Zone : ....

(ii) Percentage passing 75 mic sieve : ....

(iii) Fineness Modulus of aggregate : Sum of cumulative percentages / 100

(iv) Nature of grading : ....

Continuous : ....

Sizes in excess or missing : ....

Coarser grading/finer grading : ....

(v) Recommendation for improvement : ....

(b) Sieve analysis results for coarse aggregate

IS sieve size	Weight of aggregate retained			Total weight				gm
	I	II	III	Sum	Average sum	percent retained	Cumulative % retained	percent passing
1	2	3	4	5	6	7	8	9
40 mm								
20 mm								
16 mm								
10 mm								
4.75 mm								
2.36 mm								
Pan								

Conclusion

(i) Size : Single size/graded......................mm

(ii) Fineness modulus of coarse aggregate : $\frac{\text{Sum of cumulative percentage}}{100}$

(iii) Nature of grading

(iv) Continuity :

Size in excess/missing :

Coarse grading/finer grading :

Whether within limits :

(v) Shape :

(vi) Recommendation for improvement :

(c) Sieve analysis results for all-in-aggregate

IS sieve size	Weight of aggregate retained			Total weight				gm
	I	II	III	Sum	Average sum	Percent retained	Cumulative % retained	Percent passing
1	2	3	4	5	6	7	8	9
80 mm								
63 mm								
40 mm								
20 mm								
16 mm								
10 mm								
4.75 mm								
2.36 mm								
1.18 mm								
600 mic								
300 mic								
150 mic								
75 mic								
Pan								

Conclusion

(i) Nominal size ....mm :

(ii) Shape : ....

(iii) Nature of grading :

(iv) Continuity :

Size in excess/missing :

Coarse/finer/average grading :

Whether within limits :

(v) Recommended suggestions for improvement :

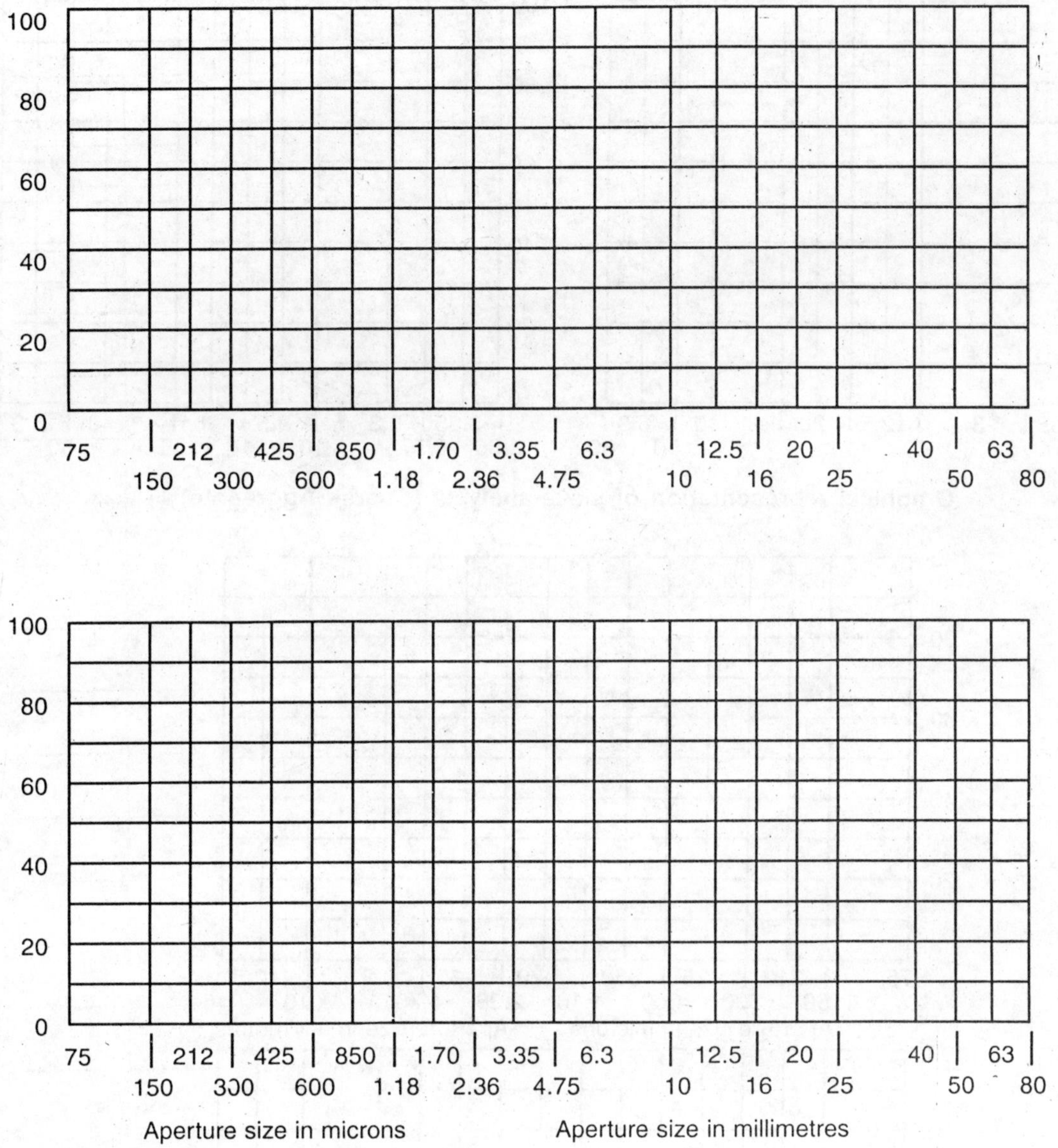

Graphical representation of sieve-analysis (all-in-aggregate)

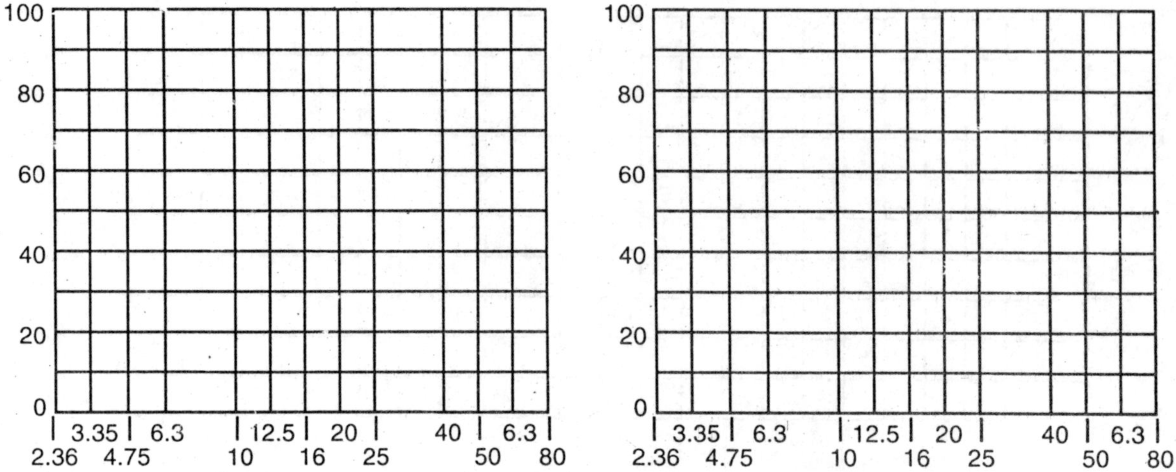

Graphical representation of sieve-analysis (Coarse Aggregate)

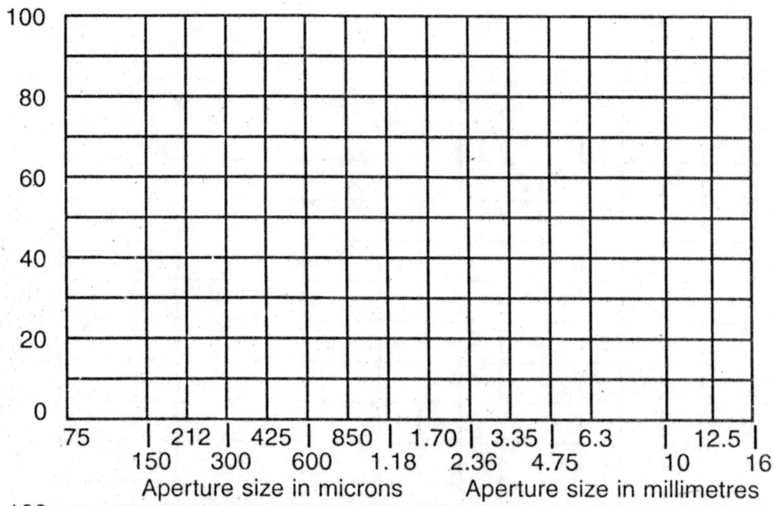

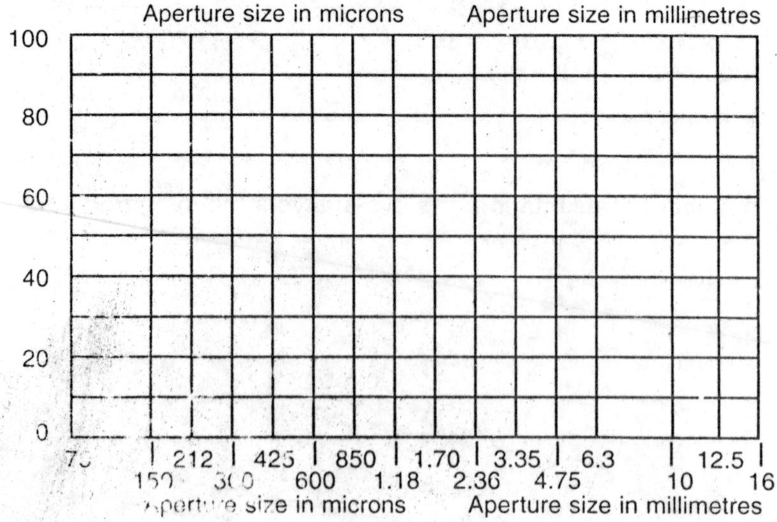

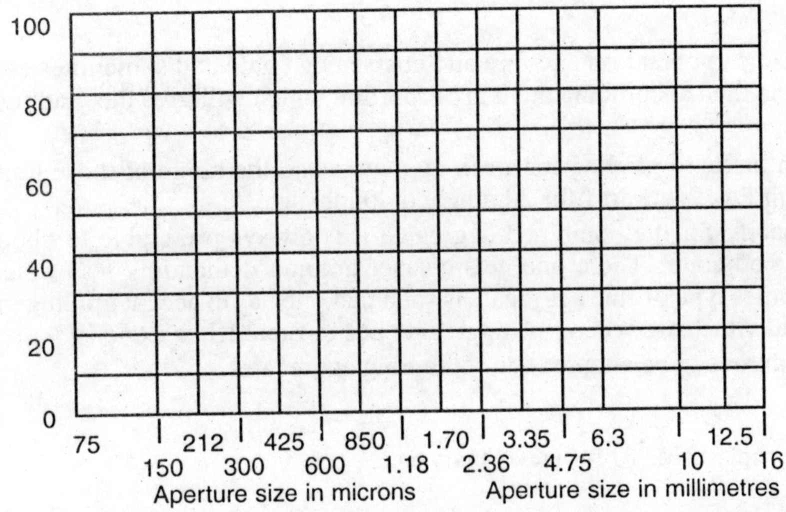

Graphical representation of sieve-analysis (Fine Aggregate)

Questions for discussion

1. When do you call aggregate as graded aggregate, single size aggregate and all-in-aggregate?
2. What is gap grading? How is it different from continuous grading?
3. How does the aggregate with one size of particles in excess of other sizes affect the workability?
4. Why is graded aggregate preferable to single size aggregate?
5. Why has ISI specified wide limits for grading instead of specifying ideal gradings?
6. What do you understand when a particular particle is said to be 20 mm, 40 mm, 10 mm size?
7. Give all sieve designation numbers normally used for sieve analysis of aggregate for concrete from the coarse aggregate?
8. How do you define fine aggregate and coarse aggregate?
9. What do you understand by representative sample of aggregate?
10. Why should sieves not be surcharged during sieving?
11. Why should aggregate be dried before sieving?

EXPERIMENT 11

METHOD OF COMBINING TWO OR MORE AGGREGATES TO GIVE A GRADING APPROXIMATE TO THE ONE REQUIRED

Statement of Problem

Given two sizes of coarse aggregates, 40 mm single size and 20 mm single size and fine aggregates whose sieve analysis results are known :

(i) Combine coarse aggregate to give graded aggregate 40 mm size.
(ii) Combine this graded aggregate with fine aggregate to give good all-in aggregate grading.

Theory

The modern practice is to obtain on site fine and coarse aggregate and sometimes each of these in two or three sizes and then to combine them in proportion which will give the grading approximate to one required.

IS 383 has given limits of grading for all-in-aggregate and the range of those limits have been shown graphically in Fig. 2.4, Part I for 20 mm and 40 mm size.

To choose or specify for the combined aggregate it is always preferable to choose a grading which lies in the 'good zone'. There one gets coarser grading demanding less water for a given workability. Sometimes type of fine aggregate is such that one has to accept grading in 'permissible zone'. This is normally the case when fine aggregate lies in Zone III or Zone IV.

Following procedure is to be adopted for combining aggregates.

Apparatus

As required in Experiment No. 10 for sieve analysis.

Procedure

When only two sizes viz. C.A. & F.A. are available.

Analytical method

(a) Enter the results of sieve analysis of each aggregate in the Part I of the observation sheet.
(b) Choose specified grading from the graph and enter the percentage passing in the Part II of the observation sheet.
(c) Choose the proportions of fine aggregate to coarse aggregate suitably. Proportion of F.A. to total aggregate normally lies between 33% to 45%. Values of percentage passing through 2.36 mm and 600 micron gives the clue for choosing the percentage of the F.A. to give the value required in specified grading and fill in the remaining columns in Part III and IV. Sample calculations are reproduced in the table below:

Analytical method

Sieve size	*Percentage passing*			*Percentage of different fractions*					
	Grading of aggregate		*Stipulated combined grading*	*I trial*			*II trial*		
	I C.A.	*II* F.A.		*I* 60%	*II* 40%	*Combined* 100%	*I*	*II*	*Combined*
80	—	—	—	—	—	—			
40	100	100	100	60	40	100			
20	97	100	100	58	40	100			
10	40	100	65	24	40	64			
4.75	5	98	42	3	39	42			
2.36	0	90	35	0	36	36			
1.18	0	75	28	0	30	30			
600	0	45	20	0	18	18			
300	0	20	7	0	8	8			
150	0	2	0	0	0.8	0.8			
75	0	0	0	0	0	0			

Note : In the above example the 1st trial gives the combined grading similar to the stipulated grading. If one does not get it, then more trials are required to be made. With more practice, two are usually enough.

Graphical method

The graphical method requires a graph sheet as shown in the Fig. 9.2. Steps are as follows :

(a) The grading of the fine aggregate is marked off on the left-hand vertical axis of the square.

(b) The grading of the coarse aggregate is marked off on the right hand vertical axis of the square.

(c) Every point indicating the percentage of any particular sieve number on the left hand vertical axis should be joined to the corresponding percentage of the same sieve size on right hand axis.

Note : Percentage of combined grading is read along A-A. Points of intersection of A-A with slopping line 1, 2, 3 etc. give percentage as indicated.

If the sieve does not exist on any of the axis then the line should be joined to the corresponding percentage of the same sieve size on the right hand axis. If the sieve size does not exist on any of the vertical axis, then the line should be joined to the 100% point on the left hand axis and 0% point on the right hand side.

(d) If coarse aggregate and fine aggregate are to be combined, then 4.75 mm sieve-size should be taken as criterion. Locate the point showing the percentage of this fraction in the specified

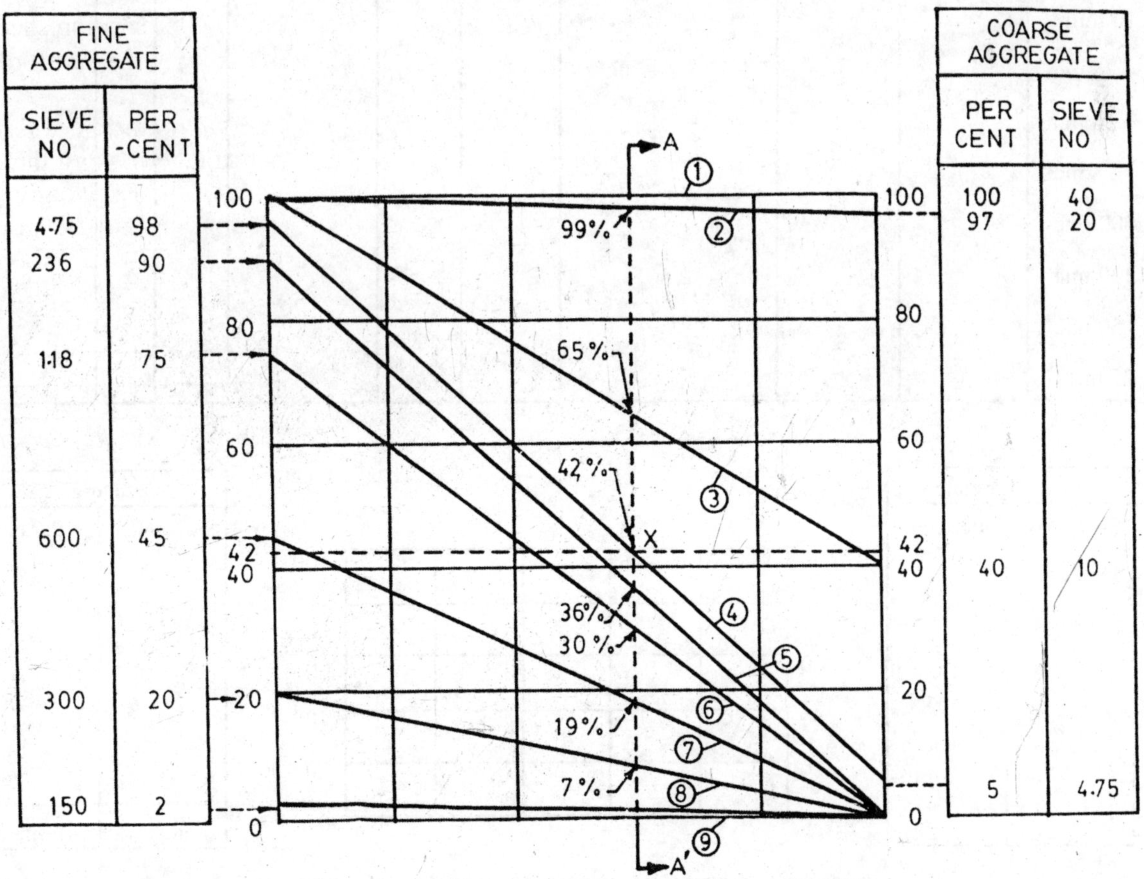

Fig. 9.2. Graphical method of combined two sizes of aggregate to obtain stipulated grading.

grading and draw a horizontal line through this point (in this case 42%). Let this line intersect the sloping line joining percentage fraction of 4.75 mm size sieve at X.

(e) Draw vertical line AA through X intersecting line (1), (2), (3), (4) etc.

(f) Measure the corresponding percentage indicated by these points of intersection, and they will give the corresponding percentages.

Observation table

IS Sieve size	*Percentage passing*			*Percentage of different fractions*					
	Grading of aggregate		*Stipulated combined grading*	*I trial*			*II trial*		
	I C.A.	*II F.A.*		*I 60%*	*II 40%*	*Combined 100%*	*I*	*II*	*Combined*
80 mm									
63 mm									
40 mm									
20 mm									
10 mm									
4.75 mm									
2.36 mm									
1.18 mm									
600 mic									
300 mic									
150 mic									
75 mic									

Fine aggregate	
Sieve No.	Per cent

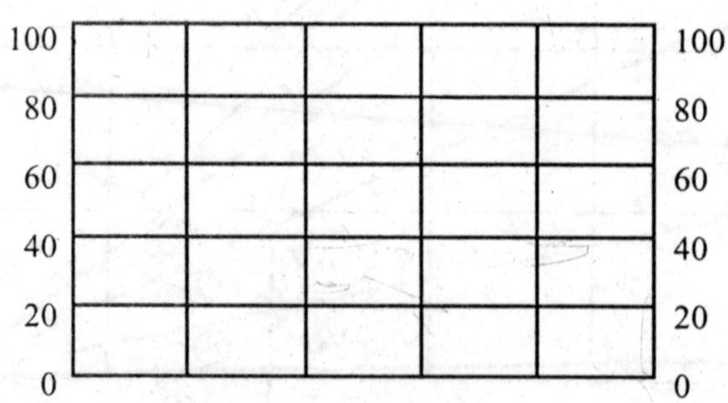

Coarse aggregate	
Sieve No.	Per cent

Graphical method of combining two size of aggregate

Questions for discussion

1. What is the difference between the grading which follows the grading curve lying within 'good zone' and 'permissible zone'?
2. Why is coarser grading preferable to finer grading?
3. Modern tendency is to obtain aggregates at site in two or three sizes. Why?

EXPERIMENT 12

METHOD OF DETERMINING AGGREGATE CRUSHING VALUE

Theory

The '*Aggregate crushing value*' gives a relative measure of the resistance of an aggregate to crushing under a gradually applied compressive load. It is the percentage by weight of the crushed (or finer) material obtained when the test aggregates are subjected to a specified load under standardized conditions, and is a numerical index of the strength of the aggregate used in road construction. Aggregates with lower crushing value indicate a lower crushed fraction under load and would give a longer service life to the road and hence a more economical performance. Weaker aggregates if used would get crushed under traffic loads, and produce smaller pieces not coated with binder and these would be easily displaced or loosened out resulting in loss of the surface/layer. In short the aggregates used in road construction must be strong enough to withstand crushing under roller and traffic.

Objectives

(i) To determine crushing value of given road aggregates.
(ii) To assess suitability of aggregates for use in different types of road pavement.

Apparatus (Fig. 9.3)

The apparatus for the standard aggregate crushing test as per IS : 2386 (Part IV) consists of the following :

(i) The test mould - a 15.2 cm diameter open-ended steel cylinder with square base plate; plunger having a piston of diameter 15 cm, with a hole provided across the stem of the plunger so that a rod could be inserted for lifting or placing the plunger in the cylinder.
(ii) A straight metal tamping rod of circular cross-section 16 mm in diameter and 45 to 60 cm long, rounded at one end.
(iii) A balance of capacity 5 kg, readable and accurate upto 1 gm.
(iv) IS sieves of sizes 12.5 mm, 10 mm, 2.36 mm.
(v) A compression testing machine capable of applying load upto 40 tonnes at a uniform rate of 4 tonnes per minute.
(vi) A cylindrical measure having internal diameter of 11.5 cm and height 18 cm.

Procedure

The aggregate sample

The material for the standard test consists of aggregate sized 10.0 mm to 12.5 mm. The aggregates should be in surface dry condition before testing. The aggregates may be dried by heating at 100°-110°C for not more than 4 hours and cooled to room temperature before testing, if necessary.

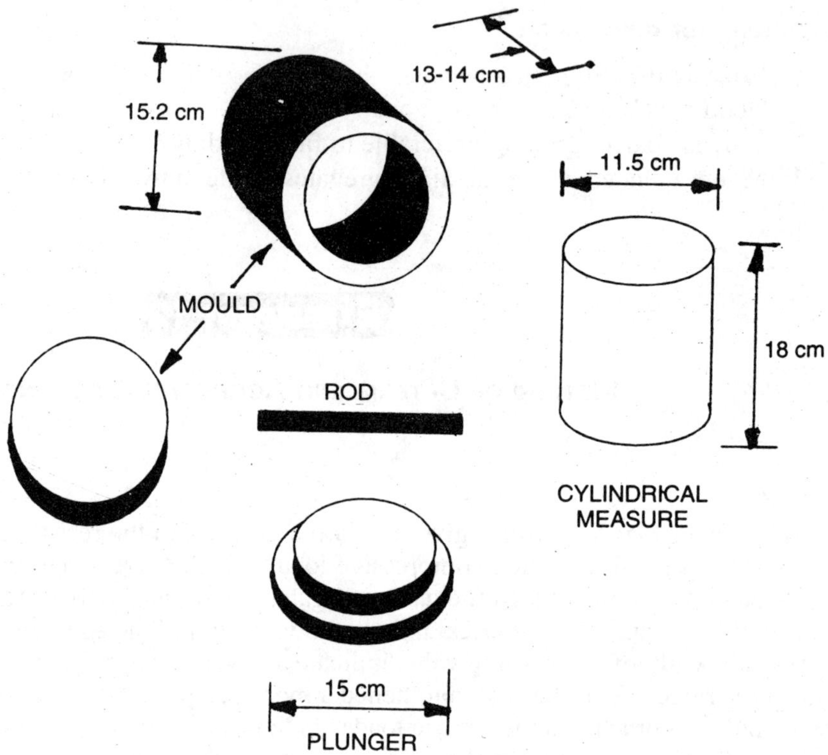

Fig. 9.3. Aggregate crushing value test apparatus.

(i) Sieve the material through 12.5 mm and 10 mm IS sieves. The aggregates passing through 12.5 mm sieve and retained on 10.0 mm sieve comprise the test material.
(ii) Take about 3.25 kg of this material.
(iii) Pour the aggregates to fill about just more than 1/3rd depth of the measuring cylinder.
(iv) Compact the material by giving 25 gentle blows with the rounded end of tamping rod.
(v) Add two more layers in similar manner, such that the cylinder is full.
(vi) Remove the excess material with a straight edge. The quantity contained in the measuring cylinder is that amount of aggregates which will be used to prepare the test specimen.
(vii) Empty the cylinder and weigh the aggregates, accurate upto 1 gm.
(viii) Transfer the whole of this weighed quantity to the test mould by filling it in three layers in the same manner as for cylindrical measure. The total depth of the sample is then about 10 cms and the surface a little below the top of mould.
(ix) Level off the surface and place the plunger over it so that it rests horizontally on the surface of the aggregates.
(x) Place this assembly on the pedestal of compression testing machine.
(xi) Apply the load at a uniform rate of 4 tonnes per minute until the total applied load is 40 tonnes.
(xii) Release the load.
(xiii) Take the aggregate out of cylinder and sieve them through 2.36 mm IS sieve. Weigh this fraction passing through it to an accuracy of 0.1 gm. This fraction is a measure of loss of material due to crushing.

(xiv) Note down the observations in the proforma and compute the aggregate crushing value. The mean of two observations, rounded to nearest whole number is reported as '*Aggregate crushing value*'.

Test with non-standard sizes of aggregates

If required, or if the standard size is not available, test may be carried out with different gradings. However specifications will be different for such cases and should conform to those given in the following table.

Aggregate size		*Quantity of material, apparatus and procedure*	*IS sieve for separating fines*
Passing through mm	*Retained on mm*		
25	20	Standard	4.75
20	12.5	Standard	3.35
10	6.3	Test cylinder—7.5 cm diameter	1.70 mm
6.3	4.75	Cylindrical measure—6.0 cm diameter and 9 cm height Tamping rod—8 mm diameter and 30 cm length	1.18 mm
4.75 3.35	3.35 2.36	Depth of material in test cylinder after tamping—5 cm Total load applied—10 tonnes uniformly in 10 minutes Quantity of material—1 kg for two samples	850 microns 600 microns

Precautions

(i) The plunger should be placed centrally and rest directly on the aggregates. Care should be taken that it does not touch the walls of the cylinder so as to ensure that the entire load is transferred onto the aggregates.

(ii) In the operation of sieving the aggregates through 2.36 mm sieve and weighing, care should be taken to avoid loss of fines. The sum of weights of fractions retained and passing the sieve should not differ from the original weight of the specimen by more than 1 gm.

(iii) The tamping should be done properly by gently dropping the tamping rod and not by hammering action. Also the tamping should be uniform over the surface of the aggregates taking care that the tamping rod does not frequently strike against the walls of the mould.

Record of observations

	Sample I	Sample II
Total weight of dry sample taken = W_1 gm		
Weight of portion passing 2.36 mm sieve = W_2 gm		
Aggregate crushing value (per cent) = $\frac{W_2}{W_1} \times 100$		

Mean = ; Aggregate crushing value =

Note : The mean of the two results to the nearest whole number is reported as the '*Aggregate crushing value*' of the material).

Interpretation of results

The suitability of aggregate is adjudged, dependent upon its proposed use in the pavement layers. The table given below lays down specified limits of per cent aggregate crushing value, for different types of road construction.

Type of road construction	*Aggregate crushing value Not more than*
1. Flexible pavements	
(a) Soling	50
(b) Water-bound-macadam	40
(c) Bituminous macadam	40
(d) Bituminous surface-dressing or thin permix carpet	30
(e) Dense-mix carpet	30
2. Rigid pavements	
(a) Other than wearing course	45
(b) Surface or wearing course	30

Questions for discussion

1. Which property of aggregates is measured by this test?
2. How is aggregate crushing value expressed?
3. What variation in the results is expected if the smaller size aggregate are used in test?
4. Should the aggregates having high ACV be considered good for road construction?
5. What are the uses of determining ACV?

EXPERIMENT 13

METHOD OF DETERMINING AGGREGATE IMPACT VALUE

Theory

The property of a material to resist impact is known as toughness. Due to movement of vehicles on the road the aggregates are subjected to impact resulting in their breaking down into smaller pieces. The aggregates should therefore have sufficient toughness to resist their disintegration due to impact. This characteristic is measured by impact value test. The *aggregate impact value* is a measure of resistance to sudden impact or shock, which may differ from its resistance to gradually applied compressive load.

Objectives

(i) To determine the impact value of the road aggregates;

(ii) To assess their suitability in road construction on the basis of impact value.

Apparatus (Fig. 9.4)

The apparatus of the aggregate impact value test as per IS : 2386 (Part IV) consists of :

(i) A testing machine weighing 45 to 60 kg and having a metal base with a plane lower surface of not less than 30 cm in diameter. It is supported on level and plane concrete floor of minimum 45 cm thickness. The machine should also have provisions for fixing its base.

(ii) A cylindrical steel cup of internal diameter 102 mm, depth 50 mm and minimum thickness 6.3 mm.

(iii) A metal hammer weighing 13.5 kg to 14.0 kg, the lower end is cylindrical in shape, is 50 mm long, 100.0 mm in diameter, with a 2 mm chamfer at the lower edge and case hardened. The hammer should slide freely between vertical guides and the concentric with the cup. The free fall of the hammer should be within 380 ± 5 mm.

(iv) A cylindrical metal measure having internal diameter of 75 mm and depth 50 mm for measuring aggregates.

(v) Tamping rod 10 mm in diameter and 230 mm long, rounded at one end.

(vi) A balance of capacity not less than 500 g, readable and accurate upto 0.1 g.

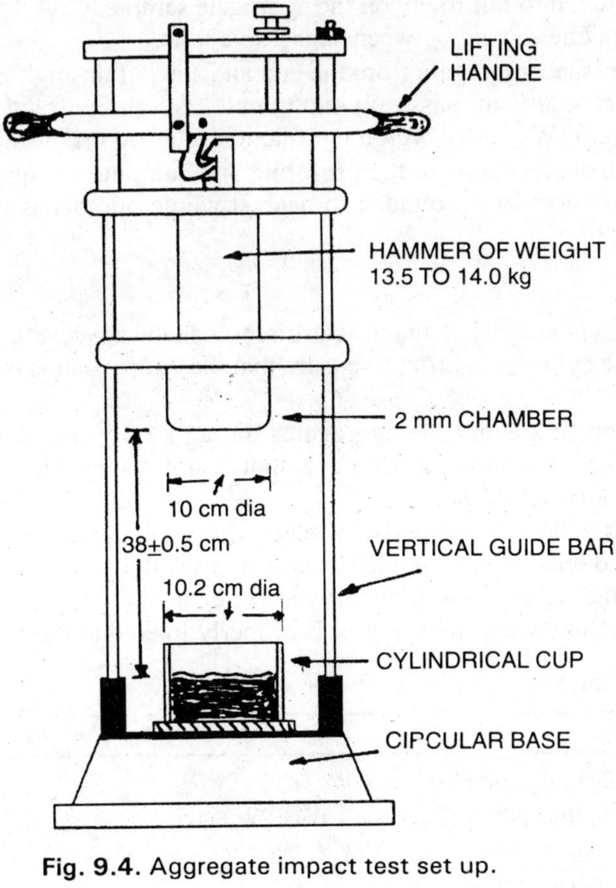

Fig. 9.4. Aggregate impact test set up.

Procedure

The test sample

It consists of aggregates sized 10.0 mm to 12.5 mm. The aggregates should be dried by heating at 100-110°C for a period of 4 hours and cooled.

(i) Sieve the material through 12.5 mm and 10.0 mm IS sieves. The aggregates passing through 12.5 mm sieve and retained on 10.0 mm sieve comprises the test material.
(ii) Pour the aggregates to fill about just 1/3rd depth of measuring cylinder.
(iii) Compact the material by giving 25 gentle blows with the rounded end of the tamping rod.
(iv) Add two more layers in similar manner, so that cylinder is full.
(v) Strike off the surplus aggregates.
(vi) Determine the net weight of the aggregates to the nearest gram (W_1).
(vii) Bring the impact machine to rest without wedging or packing upon the level plate, block or floor, so that it is rigid and the hammer guide columns are vertical.
(viii) Fix the cup firmly in position on the base of machine and place who'e of the test sample in it and compact by giving 25 gentle strokes with tamping rod.
(ix) Raise the hammer until its lower face is 380 mm above the surface of the aggregate sample in the cup and allow it to fall freely on the aggregate sample. Give 15 such blows at an interval of not less than one second between successive falls.
(x) Remove the crushed aggregate from the cup and sieve it through 2.36 mm IS sieves until no further significant amount passes in one minute. Weigh the fraction passing the sieve to an accuracy of 1 gm (W_2). Also weigh the fraction retained on the sieve.
(xi) Note down the observations in the proforma and compute the aggregate impact value. The mean of two observations, rounded to nearest whole number is reported as the '*Aggregate Impact Value*'.

Precautions

(i) Place the plunger centrally so that it falls directly on the aggregate sample and does not touch the walls of the cylinder in order to ensure that the entire load is transmitted on to the aggregates.
(ii) In the operation of sieving the aggregates through 2.36 mm sieve the sum of weights of fractions retained and passing the sieve should not differ from the original weight of the specimen by more than 1 gm.
(iii) The tamping is to be done properly by gently dropping the tamping rod and not by hammering action. Also the tamping should be uniform over the surface of the aggregate taking care that the tamping rod does not frequently strike against the walls of the mould.
(iv) Care should be taken that the hammer is properly looked at the top after the final blow.

Record of observations

	Sample I	*Sample II*
Total weight of dry sample taken = W_1 gm		
Weight of portion passing 2.36 mm sieve = W_2 gm		
Aggregate impact value = $W_2 \times 100/W_1$ per cent		
Aggregate Impact Mean value =		

Interpretation of results

Aggregate impact value is used to classify the stones in respect of their toughness property as indicated below :

Aggregate impact value	*Classification*
< 10%	Exceptionally strong
10-20%	Strong
10-30%	Satisfactory for road surfacing
> 35%	Weak for road surfacing

The Indian Road Congress has recommended the following values for different types of road construction.

S.No.	*Type of pavement*	*Maximum aggregate impact value %*
1.	Bituminous surface dressing penetration macadam, bituminous carpet concrete and cement concrete wearing courses	30
2.	Bitumen-bound-macadam, base course	35
3.	WBM base course with bitumen surfacing	40
4.	Cement concrete base course	45

Questions for discussion

1. What is meant by toughness of the aggregates?
2. How does toughness differ from compressive strength?
3. Which test simulates the field conditions better aggregate crushing value test or impact value test?
4. What are uses of determining impact value?
5. Should a good quality road aggregate give higher impact value?

CHAPTER 10

Tests on Concrete

EXPERIMENT 1

TEST FOR WORKABILITY (SLUMP TEST) TO VERIFY THE EFFECT OF WATER, FINE AGGREGAGE/COARSE AGGREGATE RATIO AND CEMENT/AGGREGATE RATIO ON SLUMP

Statement of problem

(i) For M_{15}, 220 litres of total aggregate is required to be used per bag of cement.
Slump desired is 50-100 mm.
Assume fine aggregate to coarse aggregate ratio as 1 : 2 using Zone II sand.
Keeping F.A./C.A. ratio same, calculate slump when w/c ratio is 0.50, 0.60, 0.70 and state how the slump changes with increase in water.

(ii) Keeping W/C ratio at 0.60, FA : CA = 1 : 2, observe slumps, when total aggregate is progressively reduced from 220 litres per bag to 200 litres per bag and 180 litre per bag. State how slump, cohesiveness and harshness changes with change in C/A ratio.

(iii) Keeping W/C = 0.60 (Total aggregates as 220 litres) observe slump, cohesiveness and harshness, when FA : CA ratio changes from 1 : 1, 1 : 2, 1 : 3.
State how slump, cohesiveness and harshness changes with the change in FA : CA ratio.

Theory

Workability is the ease with which concrete mix flows to the remotest corner of the form-work. In more scientific terms it is the property of concrete which determines the amount of useful internal work necessary to produce full compaction. For full compaction, concrete mix possesses three properties: Mobility, cohesiveness during movement of mix and the absence of harshness is offering smooth surface finish to trowelling. Water is the most important single factor which affects mobility, since it lubricates ingredients and reduces internal friction. But water cannot be increased indefinitely to increase mobility to avoid increase in W/C ratio. Internal friction can be reduced by decreasing the surface area by adopting coarser grading of aggregate. But too much coarseness may lead to segregation and loss of cohesiveness so essential for maintaining the homogeneity of the mix. In addition to cement content, there is a need to ensure the presence of fine particles passing through 300 micron. Harshness of concrete can be eliminated, if there is adequate proportion of mortar to fill in the voids in the coarse aggregate. Ensuring correct fine aggregate to coarse aggregate ratio is yet another factor in obtaining good workability.

There are no tests available which measure all these three properties quantitatively. Out of various tests devised to measure workability, two commonly known tools viz, slump test and compacting factor test are dealt with in this manual.

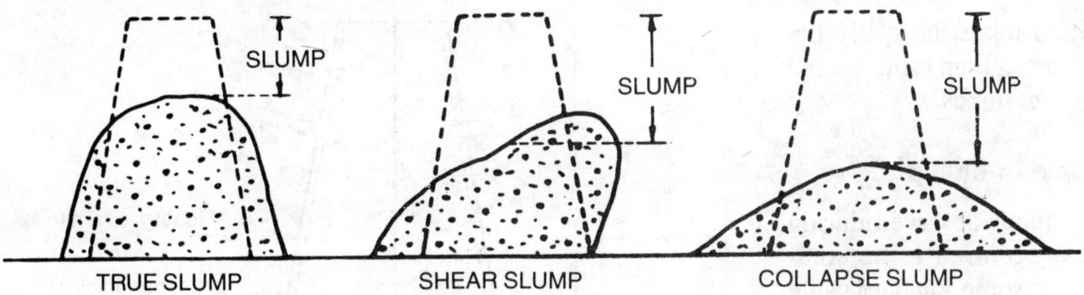

Fig. 10.1. Various types of slump.

Slump test

Slump test, the conduct of which will be described in following paragraphs, gives a measure of workability of the mix in terms of slump observed after the subsidence of a concrete mix (Fig. 10.1). One can get a fairly good idea of cohesiveness by gently tapping the platform on which cone stands. A good cohesive mix subsides further without the coarse aggregate tending to fall out of the mix during tapping. Harshness can be detected by trowelling the mix to obtain a smooth surface. Harsh concrete is normally understood and does not give smooth finish to the surface even with trowelling under pressure. Adequately sanded mix gives smooth finish with normal trowelling. Oversanded mix gives smooth finish with even light trowelling.

Limitation

Because of its simplicity in performing the experiment and its sensitivity to changes in moisture content of the successive intended identical mixes, it is widely used in the field for judging the workability.

Slump observed strictly speaking has no relation to useful internal work required for full compaction.

Also a large variation can be obtained with the same concrete. Three types of slumps are obtained :

(a) True slump,
(b) Shear slump,
(c) Collapse slump.

Collapse slump is normally obtained with lean, harsh or very wet mixes. It is difficult to measure slump when shear slump is obtained. Generally concrete giving shear or collapse slump are considered unsatisfactory for placing. Fig. 10.1 shows different types of slumps obtained.

Rich mixes normally behave better than lean, dry and very wet mixes.

Choice of slump

The slump observed during test is required to be compared to some standard value of slumps considered desirable for various types of placing and vibrating conditions. Basically, higher slump is chosen when vibration is done manually or when the sections are small and are heavily reinforced. Also greater the nominal size of aggregate, more slump is preferred. Based on the experience and experiments, various authorities such as Concrete Association of India, P.C.A. Illinois, have suggested values of slumps and are reproduced in Part I. Out of all the data given, the one given by McIntosh is more comprehensive.

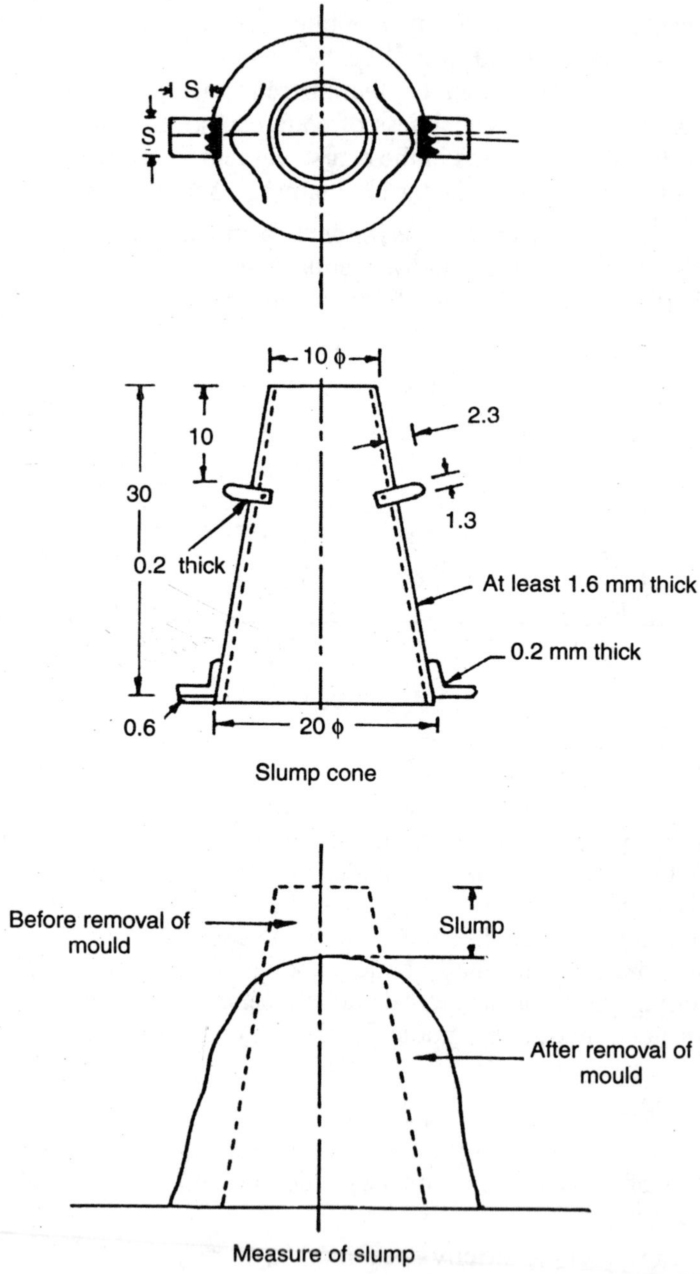

Fig. 10.2. Slump cone and measurement of slump.

The table gives the range of slumps to be aimed at. As a rule lower limits of the ranges are preferable. But in case of difficult placing conditions, it is better to aim at the higher limit of the range.

Use of slump test

The slump test is very useful on site to keep check on day-to-day, hour-to-hour variations in the materials fed into the mixes. Increase in slump may indicate unexpected increase in the moisture content of aggregate. Or it may indicate change in grading of aggregate e.g. deficiency of sand, or change in shape of the aggregate. Too-high or too-low slump gives immediate warning to the operator and enables him to remedy the situation.

For dry mixes with very low water/cement ratios, slump test gives zero slump and is not useful for high strength concrete. This test should not be used to compare workability of mixes of different proportions or types of aggregate as the results might be misleading.

Apparatus

- Mould in the form of a frustrum of a cone (Fig. 10.2)
- Tamping rod, 16 mm dia. and 600 mm long, rounded at one end
- Trough
- Trowel
- G.I. plain sheet
- Steel scale

Procedure

A. How to conduct a slump test

(a) The internal surface of the mould is thoroughly cleaned and freed from superfluous moisture and set concrete, if any, before commencing the test.

(b) The mould (see Fig. 10.2) is placed o.. a smooth, horizontal, rigid and non-absorbent surface, such as carefully levelled metal plate.

(c) The mould is held firmly in place before the concrete is filled in.

(d) Concrete under test is filled in the mould in four layers and each layer is approximately one quarter of the height of the mould. Each layer is tamped with 25 strokes of the round end of the tamping bar. The stroke should be distributed over the entire area of the mould.

(e) After the top layer has been rodded, the concrete should be struck off, levelled with a trowel or the tamping rod so that the mould is exactly filled. All mortar which have leaked out between the mould and the base plate is cleaned away.

(f) The mould is immediately raised from the concrete slowly and carefully in a vertical direction. This allows the concrete to subside and the slump is measured immediately by determining the difference between the height of the mould and that of the highest point of the specimen being tested in mm.

B. To verify the effects of FA/CA ratio, total aggregate/cement ratio and w/c ratio on slump

(a) (i) Assume suitable proportion of ingredients say 1 : 2 : 4 by volume or by weight.

(ii) Calculate ingredients for 3 kg batch of cement.

(iii) Make three trials by adding different *quantities* of water and measure slump and observe workability, harshness and cohesiveness.

(iv) It will be found that more the amount of water used, greater is the slump. Conversely in field, sudden increase in slump indicates increase in water contents if other factors remain constant and water has to be changed accordingly.

(b) Given constant W/C ratio and constant total aggregate cement ratio, observe effect of F.A./C.A. ratio on slump.

(i) Assume (suitable) proportions.

(ii) Calculate the ingredients for 3 kg batch of cement.

(iii) Make three trials by keeping W/C and total aggregate/cement ratio same, but by varying F.A./C.A. ratio and observe slump, harshness and cohesiveness.

(iv) It will be observed that deficiency in F.A. tends to segregation and harshness of concrete and more percentage of F.A. requires more water for obtaining desired slump.

(v) In field, if slump test indicates segregation and tendency to harshness it means that the aggregate is deficient in fines. Or if more water is required for desired slump, it means there is excess of fine aggregate.

(c) Given constant W/C ratio and F.A./C.A. ratio, observe the effect of A/C ratio on slump :

(i) Select suitable proportion by volume or by weight.

(ii) Calculate ingredient for 3 kg batch of cement.

(iii) Keeping W/C constant and F.A./C.A. ratio constant, make three trials by varying total quantity of aggregate.

(iv) Measure slump in each case and observe workability.

(v) It will be observed that smaller the A/C ratio, greater is the slump. In field, if the slump test starts indicating greater slump, check on batching has to be done to verify that proper quantities are added as per design.

Precautions

(a) Test shall be carried out at a place free from vibrations or shocks and within a period of two minutes after mixing if it is a field test. For laboratory test, reliable results corresponding to site conditions can be had if slump test is carried out within 10 minutes of mixing.

(b) If slump collapses or shears off laterally, test may be repeated and if again same results are obtained, the fact should be recorded and slump measured.

(c) For laboratory test, it is better to convert all quantities into weight, by multiplying volume by bulk densities. In absence of exact data, assume bulk densities from 1.5 to 1.6 for F.A. and 1.6 for C.A.

Observation table

(a) To verify the effect of water on slump, when other factors such as FA : CA ratio and aggregate/cement ratio are kept constant.

Particulars	*Cement*	*Total agg.*	*FA : CA*	*F.A.*	*C.A.*	*Water added*	*W/C ratio*	*Slump*	*Work-ability*
Initial proportions by vol/bag of cement									
Laboratory trial batch per 3 kg of cement by vol.									
Proportions by weight (multiply by bulk density)									
I trial II trial III trial									

Conclusions :

(b) To verify the effect of FA : CA ratio on workability when other factors such as aggregate : cement ratio and W/C ratio are kept constant.

Particulars	*Cement*	*Total agg.*	*FA : CA*	*F.A.*	*C.A.*	*Water added*	*W/C ratio*	*Slump*	*Work-ability*
Initial proportions by vol/bag of cement									
Laboratory trial batch per 3 kg of cement by vol.									
Proportions by weight (multiply by bulk density)									
I trial II trial III trial									

Conclusions

(c) To verify the effect of aggregate : cement ratio on slump for given W/C ratio and given FA: CA ratio.

Particulars	*Cement*	*Total agg.*	*FA : CA*	*F.A.*	*C.A.*	*Water added*	*W/C ratio*	*Slump*	*Work-ability*
Initial proportions by vol/bag of cement									
Laboratory trial batch per 3 kg of cement by vol.									
Proportions by weight (multiply by bulk density)									
I trial II trial III trial									

Conclusions :

Questions for discussion

1. In what way do the following factors affect slump?
   (a) Water content for the given proportion of ingredients and given aggregate/cement ratio.
   (b) Proportion of F.A. to C.A. for a given W/C ratio and A/C ratio.
   (c) A/C ratio for a given W/C ratio and F.A./C.A. ratio.
2. During the course of tests in field, if there is an appreciable change in the values of slump, what are the possible conclusions?
3. When slump test indicates a tendency to segregation and harshness what does it indicate? How will you correct the situation?
4. Is slump test useful for mixes with very low water cement ratio? If not why?
5. What value of slump would you suggest for following :
   (a) Bridge deck
   (b) Heavy abutment wall
   (c) Concrete pavement
   (d) R.C.C. slab in buildings
   (e) R.C.C. columns in buildings
6. What are the basic factors which determine the value of slump in each of the above types of work?

EXPERIMENT 2

TEST FOR WORKABILITY (COMPACTING FACTOR TEST)

Statement of problem

Using drier mixes in the samples prepared for slump test, take three observations on compacting factor apparatus.

Theory

Workability is the amount of work necessary to achieve full compaction of concrete. In drier mixes, slump test does not give slump and a more sensitive method to detect the change in workability is necessary. Compacting factor test works on a principle of determining the degree of compaction achieved by a standard amount of work by allowing the concrete to fall through a standard height. The degree of compaction, called compacting factor, is measured by the density ratio, i.e. the ratio of the density actually achieved in the test to density of concrete fully compacted. Thus it is, therefore, rational method than slump test and is particularly suited to dry mixes with low slump. This is useful in laboratory testing but if possible could be conducted on site. In Part I is given the table of values of slump and compacting factor suggested by McIntosh which is useful.

Apparatus (Fig. 10.3)

- Compacting factor apparatus as per IS : 1199
- Two trowels

- Hand scoop
- Tamping rod
- Platform weighing machine

Procedure

(a) Place the sample of concrete gently in the upper hopper with a hand scoop. Fill up the hopper with concrete and level it upto the brim (see Fig. 10.3).
(b) Open the trap door so that the concrete falls into the lower hopper.
(c) If the concrete sticks to the sides of the hoppers, push it gently with the help of rod from top.
(d) Open the trap door of the lower hopper and allow concrete to fall into the cylinder.
(e) With the plane of the blade of the trowel in each hand and moving them simultaneously one from each side across the top of the cylinder and at the same time keeping them pressed on

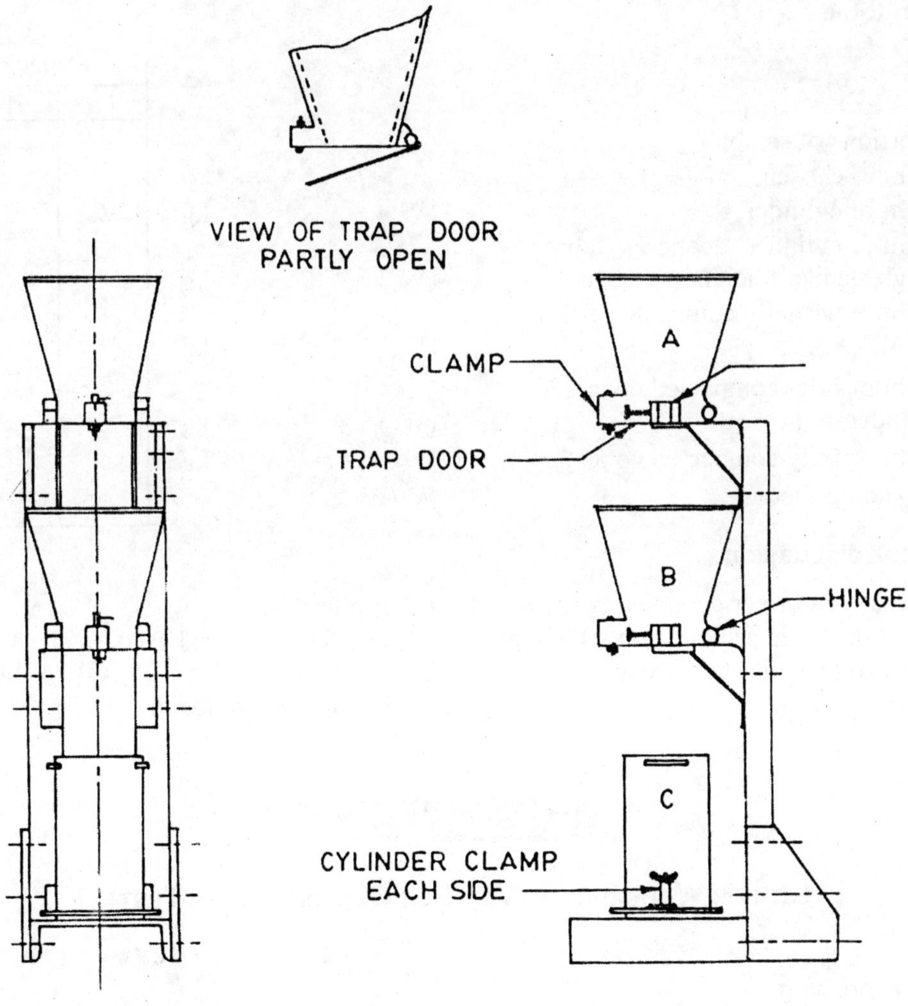

Fig. 10.3. Compacting factor apparatus.

the top edge of the cylinder, remove the excess concrete above the level of the top of the cylinder.

(f) Clean the outside of the cylinder.

(g) Determine the weight of the cylinder to the nearest 10 g.

(h) Refill the cylinder from the same samples of concrete in layers approximately 5 cm deep, layers being heavily rammed or preferably vibrated so as to obtain full compaction.

(i) Clean the outside of the cylinder and weight it again.

Precautions

To obtain strictly comparable, results, the test should be carried out at constant time interval after the mixing is completed. The convenient time for releasing the concrete from the upper hopper has been found to be 2 minutes after the completion of mixing.

Observation table

			Specimen		
			I	II	III
(a) Proportions of sample					
(b) W/C ratio chosen					
(c) Weight of cylinder	(W_1)	kg			
(d) Weight of cylinder + concrete falling through standard height	(W_2)	kg			
(e) Weight of partially compacted concrete $W_2 - W_1 =$	(W_3)	kg			
(f) Weight of fully compacted concrete + cylinder	(W_4)	kg			
(g) Weight of fully compacted concrete	$W_3 - W_1 = (W_5)$	kg			
(h) Compacting factor	W_3/W_5				

Questions for discussion

1. When do you prefer compacting factor test to slump test?
2. What is the basic principle on which the compacting factor test is based?
3. How is compacting factor expressed?
4. What precautions would you take to obtain strictly comparable results?

EXPERIMENT 3

TEST FOR WORKABILITY (VEE BEE CONSISTOMETER TEST)

Statement of problem

To determine the workability of freshly mixed concrete by Vee Bee Consistometer method.

Theory

This test is performed to determine the workability of very dry mixes. This test is developed using the principle of remoulding test with certain modifications in the apparatus and compaction by vibration instead of jolting. The apparatus was first developed by V. Bahmer of Sweden. In this test, time is measured in seconds for transforming, by vibration a concrete specimen in the shape of a conical frustum into a cylinder.

Apparatus

- The Vee Bee Consistometer
- Two trowels
- Hand scoop
- Standard tamping rod
- Stopwatch
- Weighing machine
- Scale

Procedure

(a) Prepare the concrete mix with the designed mix proportions.
(b) Fill the conical mould by placing it in the cylindrical pot.
(c) The filling of the conical mould is carried out as is done in the slump test.
(d) The glass disc attached to the swivel arm is moved and adjusted so that it just touches the top of the slump cone concrete in the pot.
(e) Now swing back the glass disc to clear the cone.
(f) The cone shall then be lifted up and the slump noted on the graduated rod by lowering the glass disc to touch the top of the concrete cone.
(g) At this point, switch on the electrical vibrator and simultaneously start the stop watch.
(h) The vibration shall then be continued to level off the concrete surface and the time taken for this (conversion of conical to cylindrical mould) shall be noted with a stop watch in seconds.

Precautions

(a) The remoulding is assumed to be complete when the glass plate rider is completely covered with concrete and all cavities in the surface of the concrete have disappeared. This is judged visually and may introduce error in establishing the end point.
(b) The slump test to be performed before giving vibrations to the table, should be performed as per the standard practice.
(c) The electric switch and stopwatch should be started at the same instant.

Observation table

S. No.	*Proportions of the mix*					*Vee bee time (seconds)*	*Slump (mm)*
	W	*C*	*FA*	*CA*	*W/C*		
1							
2							
3							

EXPERIMENT 4

TO DETERMINE COMPRESSIVE STRENGTH OF CONCRETE BY CUBE AND CYLINDER TEST

Statement of problem

From samples prepared for various phases i.e. (a), (b) and (c) of slump test, select one sample from each and fill in cube mould and cylindrical mould from each, thus preparing total 6 samples, and cure them for 28 days in prescribed manner.

Determine compressive strength for each specimen.

Theory

Concrete is primarily strong in compression and in actual construction, the concrete is used in compression. Concrete besides strong in compression is also good in other qualities. Higher the compressive strength, better is the durability. Bond strength also improves with the increase in compressive strength and is important in R.C.C. work. Compressive strength also indicates extent of control exercised during construction. Resistance to abrasion and volume stability improves with the compressive strength. Test for compressive strength is, therefore, very important in quality control of concrete. Preparation and conduct of compressive strength is comparatively easy and give more consistent results than tensile strength or flexure strength. This test for determining compressive strength of concrete has, therefore attained maximum importance. For acceptance criteria, please refer to IS : 456 or Part I of this manual. Generally 15 cm cubes are used for testing at 28 days but 10 cm size is also used if the size of the aggregate is not greater than 20 mm.

Cylinder used is 15 cm dia and 30 cm high. Whenever cylinders are used for compressive strength results, the cube strength can be calculated with the following formula : Minimum cylinder compressive strength required = 0.8 compressive strength specified for 15 cm cube.

Apparatus

- Cube moulds 10 cm size and 15 cm size as per IS : 516
- Cylindrical mould 15 cm dia. × 30 cm high as per IS : 516
- Trowels
- G.I. sheet for mixing
- Tamping rod 16 mm dia. and 600 mm long and bullet pointed at the lower end
- Glass plate thicker than 6.5 mm or machined metal plate 13 mm thickness and of dimensions greater than 17.5 mm
- 100 ton compression testing machine

Procedure

(a) Fill concrete into the mould in layers approximately 5 cm deep by moving the scoop around the top edge of the mould as the concrete slides from it, in order to ensure a symmetrical distribution of the concrete within the mould.

(b) *Compaction* : If compaction is done by hand, tamp the concrete with the standard rod, strokes being uniformly distributed over the cross-section of the mould. For 15 cm cube, number of strokes should not be less than 35 per layer and 25 strokes for 10 cm cubes. For cylindrical specimens number of strokes shall not be less than 30 per layer. Strokes should penetrate into the underlying layer. Tamp the sides of the mould to close the voids left by tamping bars.

(c) If compaction is done by vibration, then each layer is compacted by means of suitable vibrating hammer or vibrator or vibrating table. Mode and quantum of vibration of laboratory specimen shall be as nearly the same as those adopted in actual operations.

(d) *Capping* : Cylindrical specimens are capped with a thick layer of neat cement generally two or three hours after moulding operation. Cap shall be formed by a glass plate or metal plate. Work the plate on the mould till its lower surface rests on the top of the mould.

The cement for capping shall be mixed to a stiff paste for about two hours before it is to be used in order to avoid tendency of the cap to shrink. Adhesion of the paste to the capping plate can be avoided by coating the plate with a thin coat of oil or grease.

(e) *Curing* : Store the specimen in a water bath at temperature of $27^{\circ} \pm 2^{\circ}C$ for $24 \pm ½$ hour from the time of addition of water to dry ingredients. Remove the specimen from the mould and keep it immediately submerged in clean, fresh water and keep there until taken out just prior to test. Water in which specimen is submerged shall be renewed every seven days.

(f) *Test for compressive strength* :

(i) *Age at test* : Usually testing is done after 7 days and 28 days, the days being measured from the time the water is added to the dry ingredients.

(ii) Test at least three specimens at a time.

(iii) Test specimens, after about half an hour on removal from the water, till it is in surface saturated dry condition. If the specimens are received dry, keep them in water for 24 hours before testing.

(iv) Note down the dimensions of the specimens nearest to 0.2 mm and also note down their weight.

(g) *Placing specimen in the machine* :

(i) Place the specimen in such a manner that the load shall be applied to opposite sides of cubes as cast, i.e. not to the top and bottom.

(ii) Align carefully the centre of thrust of the spherically seated platen.

(iii) Apply load slowly and at the rate of 140 $kg/cm^2/min$ till the cube breaks.

(iv) Note down the maximum load and appearance of the concrete failure i.e. whether aggregate has broken or cement paste has separated from the aggregate etc.

Precautions

(i) See that the load is applied in the centre. Even a small eccentricity can cause serious deviations.

(ii) Note down the dimensions of the specimens nearest to 0.2 mm and also note down their weight.

Observation table

		Specimen I	Specimen II	Specimen III
(a) Concrete mix M100/150/200/250 :				
(b) Prescribed strength :				
(c) Identification number :				
(d) Casted on :	Date			
	Time			
(e) Tested on :	Date			
	Time			
(f) Measurements :	(a) Length (mm)			
	(b) Breadth (mm)			
	(c) Height (mm)			
(g) Area subjected to compression : A = a × b				
(h) Volume V = A × c mm^3				
(i) Weight of cube (W) kg				
(j) Unit weight of cube (W/V) kg/lt				
(k) Breaking load (P) kg	7 days			
	28 days			
(l) Compressive strength = P/A (kg/cm^2)	7 days			
	28 days			
(m) Average compressive strength (kg/cm^2)	7 days			
	28 days			
(n) Percentage deviation from average value	7 days			
	28 days			
(o) Final average compressive strength (kg/cm^2)	7 days			
	28 days			

Questions for discussion

1. Why is the compression test supposed to be most important for concrete?
2. Discuss the acceptance criterion for concrete as given in IS 456?
3. Under what circumstances 10 cm cubes are permissible?
4. What is the relation between the strengths obtained from 10 cm cube, 15 cm cube and 20 cm cube?
5. How is the cylinder strength related to the cube compressive strength?

EXPERIMENT 5

DETERMINE FLEXURE STRENGTH OF CONCRETE

Statement of problem

From samples prepared for various phases i.e. (a), (b) and (c) of slump test, select one sample from each and fill in beam mould 15 × 15 × 70 cm from each, thus preparing total 6 samples, and cure them for 28 days in prescribed manner.

Determine flexure strength for each specimen.

Theory

The knowledge of tensile strength in concrete is of value in estimating the load under which cracking will develop. The absence of cracking is of considerable importance in maintaining the continuity of a concrete structure and preventing corrosion of reinforcement.

Maximum tensile stress reached in the bottom fibre of the test beam is known as 'Modulus of Rupture' of concrete (Refer IS 456).

Apparatus (Fig. 10.4)

- Standard beam mould 15 × 15 × 70 cm for aggregate size upto 30 mm. If the largest size of the aggregate does not exceed 20 mm, the size may be 10 × 10 × 50 cm. For detailed specification, see IS : 516.
- Tamping bar
- Trowels
- Hand scoop
- Universal testing machine capable of applying load at suitable rate (4 KN/minute for 150 mm specimens and 1.8 KN/minute for 100 mm specimens)

Procedure

(a) Method of filling the mould, curing and measuring the dimensions is the same as is done for the compressive strength test.

(b) Place the specimen in the testing machine such that the load shall be applied to the uppermost surface as cast in mould, along middle third points spaced 20 cm or 13.3 cm apart (see Fig. 10.4). The axis of the specimen shall be carefully aligned with the axis of the loading device. No packing shall be used between the bearing surfaces of the specimen and the rollers.

(c) Apply load carefully without shock and increasing continuously at a rate of 400 kg/min for 15.0 cm specimen and at the rate of 180 kg/min for 10 cm specimen. The load shall be increased until the specimen fails and the maximum load applied to the specimen during the test shall be recorded.

(d) Measure the distance between the line of fracture and the nearer support 'a'. Note if 'a' is greater than 200 mm for 150 mm specimen or greater than 133 mm for 100 mm specimens.

(e) Determine the modulus of rupture (flexure strength) by formula $fb = p \times l/bd^2$, when 'a' is greater than 200 mm for 150 mm specimens (or 'a' is greater than 133 mm for 100 mm specimens) or $fb = 3p \times a/bd^2$ N/mm^2 when 'a' is less than 200 mm but more than 150 mm specimens (or 'a' is lesser than 133 mm for 100 mm specimens).

p is maximum load in N (kg).

b, d, l are width, depth and span of the beam respectively all in mm.

If 'a' is less than 170 mm for 150 mm specimen or less than 110 mm for 100 mm specimens, the results of the tests are to be discarded.

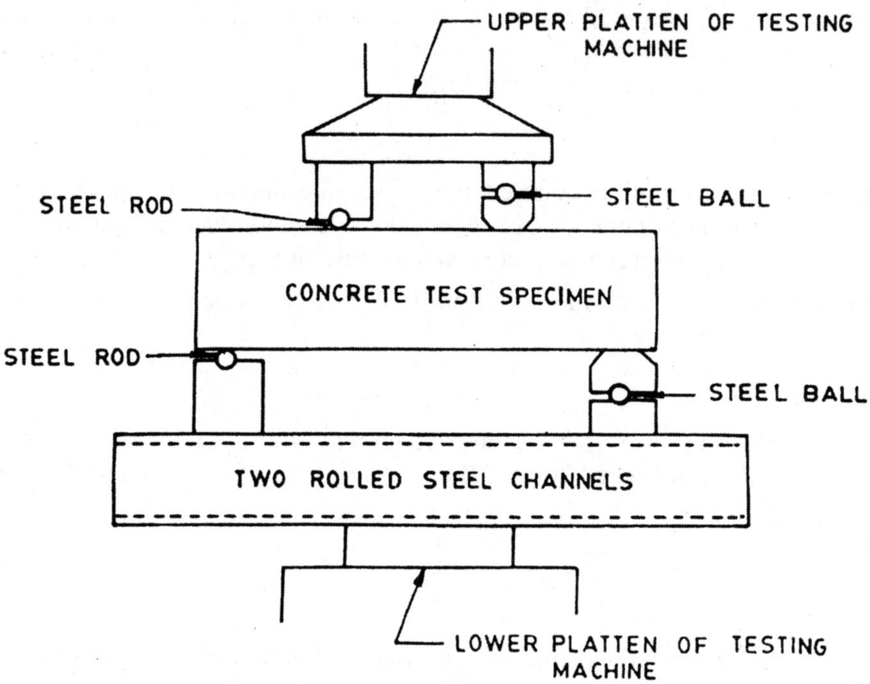

Fig. 10.4. Adaptor for testing beams in flexure.

Precautions

(a) The loading should be applied centrally and without subjecting the specimen to any torsional stresses or restraints.
(b) The axis of the specimen shall be carefully aligned with the axis of the loading device.
(c) No packing shall be used between the bearing surfaces of the specimen and the rollers.
(d) The load shall be applied without shock and increasing continuously at the specified rate.
(e) The load shall be correctly measured at the time of development of cracks.
(f) The location of cracks shall be correctly measured from the supports.

Observation table

S.No.	*Mix proportions*				*W/C ratio*	*Size of specimen*	*Maximum load N (kg)*	*Modulus of rupture*
	W	*C*	*FA*	*CA*				
1								
2								
3								
4								
5								
6								

Questions for discussion

1. In what way is the flexure test used as a criteria for acceptance of concrete?
2. What do you understand by modulus of rupture?

EXPERIMENT 6

FIELD METHOD OF DETERMINING THE PROPORTIONS OF INGREDIENTS FOR ORDINARY CONCRETE MIX

Statement of problem

Given : Prescribed mix $M_{15}/M_{20}/M_{25}$
20 mm coarse aggregate graded/angular/rounded
Fine aggregate zone I, II, III, IV

Design a mix for ordinary concrete.

Note : Even though this is a field method, it may be conducted in the laboratories by first conducting tests on 3 kg or 5 kg sample of concrete and final trial on 10 kg sample of concrete. 3 kg sample means a sample of concrete in which proportions of ingredients are based on 3 kg of cement.

Theory

There are basically two methods of designing concrete mixes :

(a) Design by nominal mix proportions (ordinary concrete).
(b) Design for maximum strength.

In the first case, the determination of the proportions of cement, aggregate and water to attain the required strength is done by adopting nominal mixes. No trial mixes are made in the laboratory. This type of concrete is called the ordinary mix.

IS 456 has given these nominal mixes which are now reproduced in Table. Ordinary concrete strength are limited to M_{25} provided strict control on W/C ratio and quality of aggregate is maintained.

Full scale field trial mixes have to be made to determine final proportion of fine aggregate to coarse aggregate and proportion of total aggregate to cement for a given water/cement ratio and workability.

Proportion of fine aggregate to coarse aggregate can be chosen for the initial trial from table of proportioning as given in part I.

Reliable results can be had if the trial mixes are made with 50 kg of cement. For demonstrating to students in the laboratory, however, 10 kg batch will be good enough.

During slump test, slump as well as the cohesiveness of the mix also should be observed by tapping gently the platform on which the cone stands. If the mix tends to segregate, it is advisable to increase the proportion of fine aggregate.

As a general rule, increase in fine aggregate is necessary for :

(a) Leaner mixes such as M_{10}, M_{15}.
(b) More workable mixes.
(c) When concrete is made with aggregate with poorly shaped particles (angular, flaky etc.).
(d) When maximum size of coarse aggregate is small.
(e) Where concrete is likely to be subjected to conditions leading to segregation such as transporting long distances or appreciable heights, or formwork permitting leakage of mortar.
(f) Where sections are small or heavily reinforced.

Since W/C ratio is to be kept constant, slump requirement can be obtained only by changing aggregate/cement ratio. Total quantity of aggregate per bag of cement mentioned in IS : 456 for nominal mixes should not be exceeded under any circumstances. If the actual slump is less than required, reduce the total aggregate. If the actual slump is more than required, reduce quantity of water only, thus reducing W/C ratio which is permissible. If the mix tends to segregate, increase proportion of F.A. Care must be taken to make adjustment to bulking and moisture content.

Apparatus

- Concrete mixer 100 litre capacity
- Slump test apparatus
- Containers of suitable capacity
- Spring balance up to 100 kg

Procedure

Stepwise procedure for manufacture of ordinary concrete

Foregoing discussions have helped us to understand the principles underlying the manufacture of concrete. Following example offers steps to be followed in field to manufacture ordinary concrete.

Example : Field determination of proportions for various ingredients of M_{15}.

Grade of concrete : M_{15}.

Maximum size of aggregate : 20 mm graded, moist.

Fine aggregate : Sieve analysis indicates zone III; squeezing test indicates that it is moderately wet.

Desired slump range : 25 mm to 50 mm; capacity of concrete mixes; 200 litres, i.e. one bag of cement to be used per batch.

Vibration : Mechanical.

Step 1 : Choose initial proportions. Referring to Table, Part I, write down the data given. Since the zone of sand is III, and maximum size of aggregate is 20 mm, the proportion of FA : CA is 1 : 3 as seen from Table.

Particulars	Cement kg	Total dry aggr.	FA/CA	F.A.	C.A.	W/C	Water
(1)	(2)	(3)	(4)	(5)	(6)	(7)	(8)
Initial assumption	50	220 lt.	1 : 3	55	165	0.64	32 lt.

Step 2 : Make adjustment for water to be added at the mixes, after examining the conditions of aggregate. Example states that the fine aggregate is moderately wet and coarse aggregate is moist.

From Table in Part I, surface moisture available is 80 litres/1000 litres of fine aggregate and 40 litres/1000 litres of coarse aggregate.

(1)	(2)	(3)	(4)	(5)	(6)	(7)
Correction for moisture :						
Moisture in FA $\frac{80}{1000} \times \frac{55}{1}$						4.4
Moisture in CA $\frac{40}{1000} \times \frac{165}{1}$						6.6
					Total water	11.0

Step 3 : Conduct test for bulking of fine aggregate and make adjustment for bulkage. If tests are not made, increase the fine aggregate by 25 per cent.

Assume in this example, that bulkage determined on site is 25 per cent.

(1)	(2)	(3)	(4)	(5)	(6)	(7)
Correction for bulkage						
@ 25% F.A. volume = $\frac{125}{100} \times \frac{55}{1} = 70$						

Step 4 : Estimate quantities after correction.

(1)	(2)	(3)	(4)	(5)	(6)	(7)
Quantities after corrections	50 kg	220	1 : 3	70	165	21.0 lts

Step 5 : Calculate the quantities required per batch according to the capacity of the mixes. For example 100 litre capacity mixer, will consume only ½ bag at a time, 200 litres will consume 1 bag of cement per batch.

In this particular case, 200 lt capacity mixer will consume 1 bag of cement per batch.

(1)	(2)	(3)	(4)	(5)	(6)	(7)
Quantities per batch	50	220	1 : 3	70 lt	165 lt	21.0 lt

Step 6 : Select suitable sizes of farmas (containers). See table in Part I.

For F.A., use 35 litre twice.

For C.A., use 35 litre thrice and 25 litre once.

Step 7 : Feed all the ingredients into the mixer. At this stage, add exactly moisture content arrived at after correction (210 lt. in this case). The concrete supervisor at this moment should look into the mix in the mixer after about 3 min. of mixing and judge consistency of the mix. If he thinks that the mix is too dry to get the desired slump, more water should be added. But the exact record of the water added should be maintained. Discharge the mix and conduct slump test at least three times by taking samples from various different places in a heap. Measure the slump and observe its cohesiveness and harshness.

To explain this example let us say it is found that water required to get the desired slump was more than the calculated (say 24 litres). Also there is a tendency to segregate which meant proportion of fine is less than required. Firstly, therefore, increase aggregate/cement ratio. Instead of 50 kg/220 litres, use 50 kg/200 litres. Secondly, there is a tendency to segregate then change FA : CA ratio to 1 : 2 instead of 1 : 3. These values are adopted solely on the judgement of the supervisor. Now repeat the previous calculations.

Particulars	*Cement kg*	*Total aggr.*	*FA : CA*	*FA*	*CA*	*W/C*	*Water*
1. Initial proportions	50	200	1 :2	67	134	0.64	32
2. Correction for moisture							
Moisture (FA) = $\frac{80}{100} \times \frac{67}{1}$							5.40
Moisture (CA) = $\frac{40}{100} \times \frac{134}{1}$							6.60
3. Correction @ 25% F.A. volume						Total	12.00
For bulkage = $\frac{125}{100} \times \frac{67}{1} = 84$							
4. Quantities after correction	50	200	1 : 2	84	134	0.40	20
5. Quantities per batch of cement	50	200	1 : 2	84	134	0.40	20
6. Selection of farma (container) : (a) For F.A. use 45 litres twice (b) For C.A. use 45 litres thrice							
7. Mix the ingredients and conduct the second trial and observe slump.							

Within two or three trials, one can find the correct proportions. Once the correct proportions are found, take at least three cubes for test specimen for compression test and also conduct unit weight test (see Experiment No. 7).

Observation table

(a) Prescribed mix : M10/15/20/25

(b) Nominal size of coarse aggregate, mm : ....

(c) Zone of fine aggregate : I/II/III/IV

(d) Proportion of FA : CA assumed initially : ....

(e) Moisture conditions :

  (i) C.A. : Moist/moderately wet/very wet

  (ii) F.A. : Moist/moderately wet/very wet

(f) Values of surface of water in aggregate
  (i) C.A. (litre/1000 litre) : ....
  (ii) F.A. (Litre/1000 litre) :
(g) If aggregate are dry, values of water absorption by aggregate :
  (i) C.A. (per cent by weight) :
  (ii) F.A. (per cent by weight) :
(h) Bulk-densities (loose-rodded)
  (i) C.A. (kg/litre) :
  (ii) F.A. (kg/litre) :
(i) Bulkage of sand (per cent) :
(j) Slump required :

S.No.	*Particulars*	*Cement kg*	*Total*	*FA : CA*	*F.A.*	*C.A.*	*Water litres*	*W/C*
1	2	3	4	5	6	7	8	9
1.	Initial assumption							
	I trial							
	II trial							
	III trial							
2.	Correction for moisture in FA $\frac{F(ii) \times Col.(5)}{1000 \times 1}$							
	I trial							
	II trial							
	III trial							
3.	Moisture in CA $\frac{f(ii) \times Col(6)}{1000}$							
	I trial							
	II trial							
	III trial							
4.	Correction for water absorption							
	ABS due to CA $\frac{g(i) \times col(6) \times h(i)}{100}$							
	I trial							
	II trial							
	III trial							
5.	ABS due to FA $\frac{g(ii) \times col\ 5 \times h(ii)}{100}$							

(Contd.)

1	2	3	4	5	6	7	8	9
	I trial							
	II trial							
	III trial							
6.	Correction for bulkage $\frac{100 + (i) \times \text{col } 5}{100}$							
	I trial							
	II trial							
	III trial							
7.	Quantities after correction :							
	I trial							
	II trial							
	III trial							
8.	Proportions per 3/5/7 kg batch of cement for lab experiment							
	I trial							
	II trial							
	III trial							
			I trial	II trial	III trial			
9.	Selection of container size							
	No. of times							
10.	Observations							
	Workability (slump - mm)							
	Water used							
	Remarks							

Notes

1. Water actually used to get desired slump may be more or less than the one calculated in column 7.
2. If water actually used is more, then reduce total aggregates, thereby reducing aggregate/cement ratio.
3. When the mix shows tendency to segregate, increase proportion of fine aggregate, taking care to see that this increase is as little as possible.
4. Make these changes and conduct II trial and so on.
5. If water actually used is less, then no changes are normally necessary.

Final choice of proportions by volume

Particulars	Cement	F.A.	C.A.	Total aggregate	W/C
Per kg of cement					
Per 50 kg of cement					

Questions for discussion

1. Define ordinary concrete and controlled mix?
2. What are the proportions specified by ISI for M_{25} grade of concrete?
3. Under what circumstances would you increase the proportion of fine aggregate?
4. Why is it necessary to keep W/c ratio constant? What is the influence of W/c ratio on strength of concrete?
5. How will you keep W/c ratio constant and still increase the workability to desired value?
6. If the concrete tends to segregate, how will you correct the mix?
7. While deciding final proportions, how will you adjust them to allow for bulkage and moisture contents?
8. What values would you choose for moisture content of absorption if the sand is very wet and coarse aggregate is perfectly dry?
9. What are the various sizes of box container (Farma) that are necessary for volume batching of concrete?

EXPERIMENT 7

DETERMINATION OF UNIT WEIGHT, YIELD PER BAG OF CEMENT, CEMENT FACTOR AND AIR CONTENT OF FRESHLY MIXED CONCRETE

Statement of problem

From the final mix proportions arrived at in Experiment 6, determine unit weight of concrete, yield per bag of cement, cement factor and air content of freshly mixed concrete.

Theory

It is always necessary to calculate the material required for specific amount of compacted concrete. It is also required to calculate absolute volume of concrete produced in a mixer. All these values are required in concrete mix design. Cement factor is very important in determining the economy of design.

Apparatus

- Balance
- Tamping bar
- Measure of 10 litres for aggregate up to 40 mm and 20 litres for aggregate over 40 mm size. For details see IS : 1199.

Procedure

(a) Find out the volume of the cylindrical measure by determining the weight of water at room temperature required to fill it so that no miniscus is present above the rim. Accurate filling of measure may be secured by use of a glass cover plate.

(b) Volume of measure is the weight of water divided by the unit weight i.e. 1 kg/litre.

(c) Fill the measure with concrete as soon as practicable after mixing.

(d) Fill the concrete in layers of approximately 5 cm deep and each layer shall be compacted either by hand or by vibration.

(e) Compacting by hand shall be done by the standard tamping bar, by tamping uniformly over the entire cross section with not less than 60 strokes per layer for 0.01 m^3 measure or 120 strokes per layer for the 0.02 m^3 measure.

(f) Compaction can also be done by electric or pneumatic hammer or by suitable vibrating table.

(g) Tap the exterior surface of the cylinder at least 10 to 15 times or until no large bubbles of the air appear on the surface of the compacted layer.

(h) After the top layer has been compacted, the surface of concrete shall be struck off level with the top of measure.

(i) Finish smoothly with flat cover plate using great care to leave the measure just level full.

(j) Clean the exterior of the measure and weigh the filled measure.

Observation table

		Specimens		
		I	II	III
(a) **Unit weight per m^3 (w) :**				
(i) Weight of water (W)				
(ii) Volume of measure (Vm)	lt.			
(iii) Weight of measure (W)	kg			
(iv) Weight of measure + cone (W_1)	kg			
(v) Weight of concrete $W_1 - W = W_2$	kg			
(vi) Unit weight of concrete $W = \frac{W_2}{V_m}$	kg/lt			
(b) **Volume of concrete per batch (V) :**				
(i) Weight of concrete per batch (N × 50) where N = No. of bags	kg			
(ii) Weight of fine aggregate (Wf) per batch (Wf)	kg			
(iii) Weight of coarse aggregate per batch (Wc)	kg			
(iv) Weight of mixing water per batch (Ww)	kg			
(v) Unit weight of concrete (W)	kg/lt			
(vi) Volume of concrete batch $V = \frac{N \times 50 + Wf + Wc + Ww}{W}$	lt.			
(vii) Average value				
(c) **Volume of concrete per batch by 'Absolute Volume' method (Va) :** $\frac{N \times 50}{3.15}$	lt.			

	Unit	Specimen I	Specimen II	Specimen III
$+ \dfrac{Wf}{\text{specific gravity of F.A.}}$	lt.			
$+ \dfrac{Wc}{\text{specific gravity of C.A.}}$	lt.			
$+ \dfrac{Ww}{1}$	lt.			
$= V_a$	lt.			
Average value	lt.			
(d) Theoretical weight of concrete on air free basis (T)				
$50\,N + Wf + Ww + Wc = Wt$	kg			
$T = \dfrac{W_t}{V_a}$	kg/lt.			
(e) Yield per bag of cement (Y) :				
Volume of concrete per batch (V)	lt.			
Number of 50 kg – bags (N)				
Yield per bag of cement $Y = \dfrac{V}{N}$	lt.			
(f) Cement factor (Number of bags/m^3 of concrete T_1)				
Yield of concrete/bag of cement (Y)	lt.			
Number of bags per m^3 of concrete $(T_1) = \dfrac{1}{Y}$ or $\dfrac{N}{V}$				
(g) Air content (A) : Percent Voids				
(i) Theoretical unit weight of concrete (T)				
(ii) Unit weight actually measured (W)				
(iii) Volume of concrete produced per batch (V)				
(iv) Absolute volume of all ingredients (V_a)				
(v) Air content (%age voids)				
$A = \dfrac{T - W}{T} \times 100$	%			
$A = \dfrac{V - V_a}{V} \times 100$	%			

Questions for discussion

1. Define yield of concrete per bag of cement, cement factor, and air content?

EXPERIMENT 8

METHOD OF MAKING, CURING AND DETERMINING COMPRESSIVE STRENGTH OF ACCELERATED - CURED CONCRETE (REFER IS : 9013)

Theory

Traditionally, quality of concrete in construction work is calculated in terms of its 28 days compressive strength. This procedure requires 28 days of moist curing before testing, which is too long a period to be of any value for either concrete construction control or applying timely corrective measures. Thus by the time the results of 28 day test, or even 7 day test are available, a considerable amount of concrete has been utilized. It is then rather too late for remedial measures if the concrete is too weak or if it is too strong. With the assistance of reliable test methods employing accelerated curing techniques, it is now possible to test the compressive strength of concrete within a short period and thereby to estimate whether it is likely to reach the specified strength at 28 days or not. The two methods of such accelerated techniques are given as follows :

(i) Warm-water method, and
(ii) Boiling-water method.

Apparatus

- Mould conforming to IS : 516
- Mould cover plate
- Curing tank (Diagrammatic layout given in Fig. 10.5).

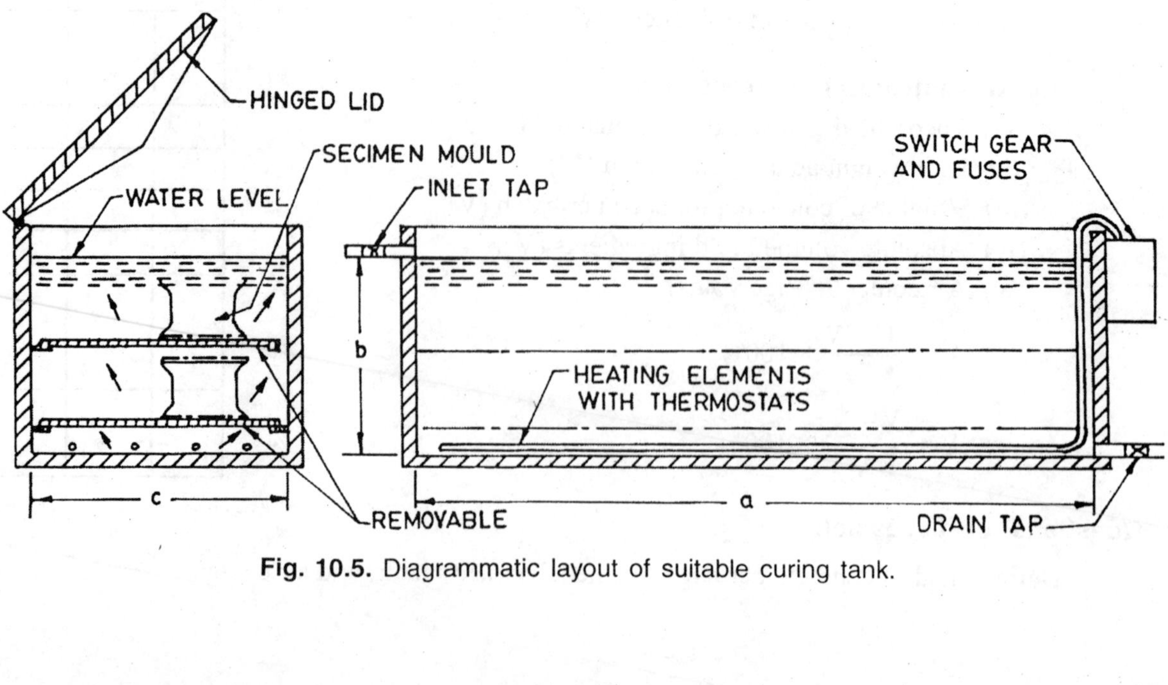

Fig. 10.5. Diagrammatic layout of suitable curing tank.

Procedure

1. Preparation of test specimens

(a) Preparation of test specimens shall be in accordance with as given in Experiment No. 4.

(b) Immediately after moulding, each specimen shall be covered with a steel plate thinly coated with mould oil to prevent adhesion of concrete.

2. Accelerated curing by warm-water method

(a) After the specimens have been made, they shall be left to stand undisturbed in their moulds in a place free from vibration at a temperature of 27 ± 2°C for at least one hour, prior to immersion in the curing tank.

(b) The time between the addition of water to the ingredients and immersion of the test specimens in the curing tank shall be at least 1½ hours but shall not exceed 3½ hours.

(c) The specimens in their moulds shall be gently lowered into the curing tank and shall remain totally immersed at 55 ± 2°C for a period of not less than 19 hours 50 minutes.

(d) The specimens shall then be removed from the water, marked for identification, removed from the moulds and immersed in the cooling tank at 27 ± 2°C before the completion of 20 hours, 10 minutes from the start of immersion in the curing tank. They shall remain in the cooling tank for a period of not less than one hour.

3. Accelerated curing by boiling water method

(a) After the specimens have been made, they shall be stored in a place free from vibration, in moist air of at least 90 per cent relative humidity and at a temperature of 27 ± 2°C for 23 hours ± 15 minutes from the time of addition of water to the ingredients.

(b) The specimens shall then be gently lowered into the curing tank and shall remain totally immersed for a period of 3½ hours ± 5 minutes. The temperature of the water in the curing tank shall be a boiling (100°C) at sea level.

(c) The temperature of water shall not drop more than 3°C after the specimens are placed and shall return to boiling within 15 minutes.

(d) After curing for 3½ hours ± 5 minutes in the curing tank, the specimen shall be removed from the boiling water, removed from the moulds and cooled by immersing in cooling tank at 27 ± 2°C for a period not less than one hour.

4. Testing

(a) The specimens shall be tested as in Experiment No. 4.

(b) In the warm-water method specimens shall be tested while still wet, not more than 2 hours from the time of immersion in the cooling tank.

(c) In the boiling water method, the age at the time of test subsequent to removal of moulds and cooling, shall be 28½ hours ± 15 minutes.

Precautions

The following precautions shall be taken :

(a) The curing tank shall be cleaned and the water removed periodically so as to prevent accumulation of deleterious material which may impair the heating or circulating system.

(b) The use of boiling water imposes the need for safety measures to prevent scalding or eye-burns, resulting from sudden escape of steam, upon opening the cover. Also care shall be exercised when immersing the specimens to avoid splashing of hot water.

(c) Strict attention shall be given to the protection and storage of the specimens during the initial period of curing.

(d) Suitable safety devices and indicators shall be provided with the set up. A separate panel or switch board shall be provided incorporating the therograph and related heating equipment controls.

Observation table

A. Warm water method

S.No.	*Size of specimen (mm)*	*Load at failure (N)*	*Compressive strength (N/mm^2)*
1			
2			
3			Average: ....

B. Boiling water method

S.No.	*Size of specimen (mm)*	*Load at failure (N)*	*Compressive strength (N/mm^2)*
1			
2			
3			Average: ....

Interpretation of results

(a) Since strength requirements in existing specifications are not based upon accelerated curing, results from this method in checking the compliance of specified strength at later ages shall be applied with great caution.

(b) The results can be used in rapid assessment of variability for process control and signalling the need for indicated adjustments.

(c) The magnitude of the strength values from strength tests is influenced by the specific combination of materials. Therefore, the use of the results from either conventional tests at any arbitrary age or those from this method shall be supported by experience or correlations developed for the existing local conditions and materials.

Correlation of results from compressive strength tests on specimens cured by normal and accelerated curing methods

Accelerated curing of concrete hastens the process of hydration of cement and as a result, a substantial proportion of the strength to be attained in 28 days under normal curing conditions is achieved within a shorter time. The rate and extent of hydration of cement under a particular curing regime depend mainly upon the chemical composition of cement, water-cement ratio and mix proportions, which are considered to be important parameters in the correlation of results from compressive strength tests on specimens cured by accelerated curing method and normal curing method.

The accelerated curing regime, in itself, is another variable in that the higher temperature employed may alter the morphology of the hydration products apart from thermally activating the chemical reactions of hydration of cement. Figures given below show typical results on the correlation of compressive strength of concrete specimens normally cured and accelerated-cured by the boiling water method and warm water method respectively. It is found that a correlation exists between the results obtained on concrete specimens, cured by accelerated method and cured by normal method, for mixes employing different materials and mix proportions. It is also found that

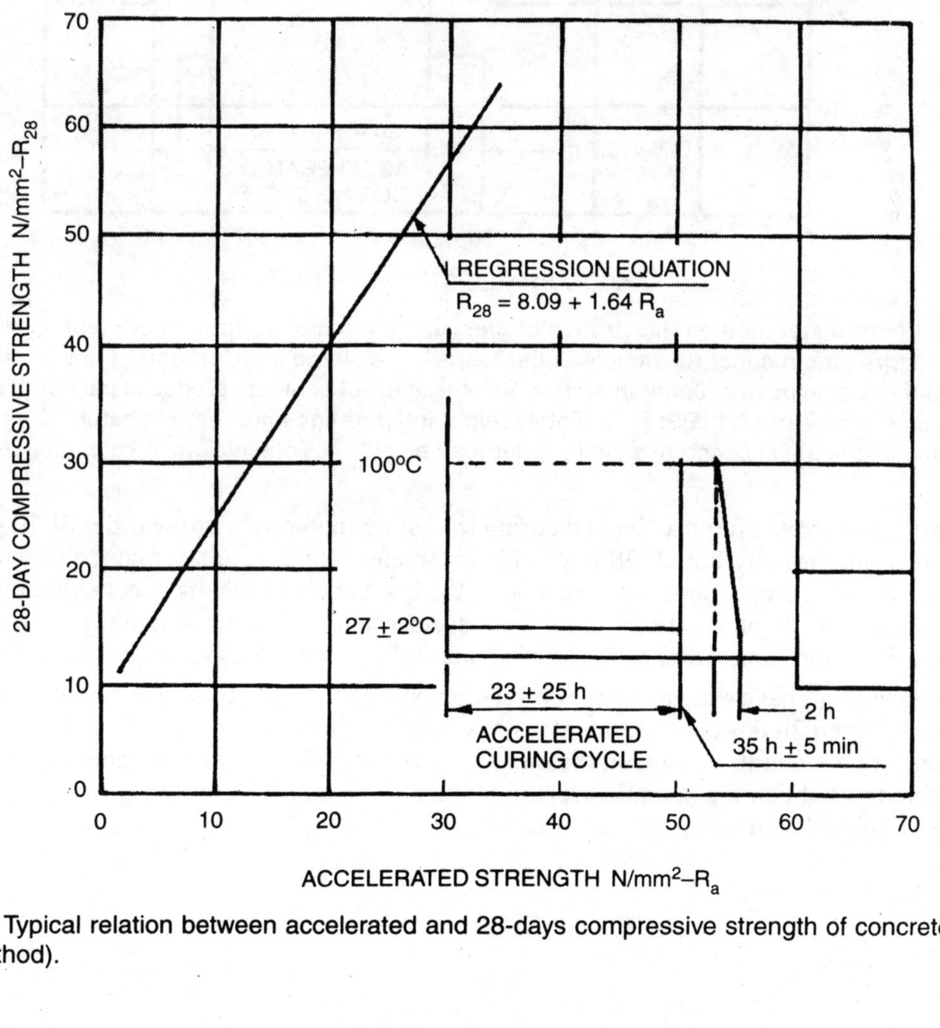

Fig. 10.6. Typical relation between accelerated and 28-days compressive strength of concrete (Boiling water method).

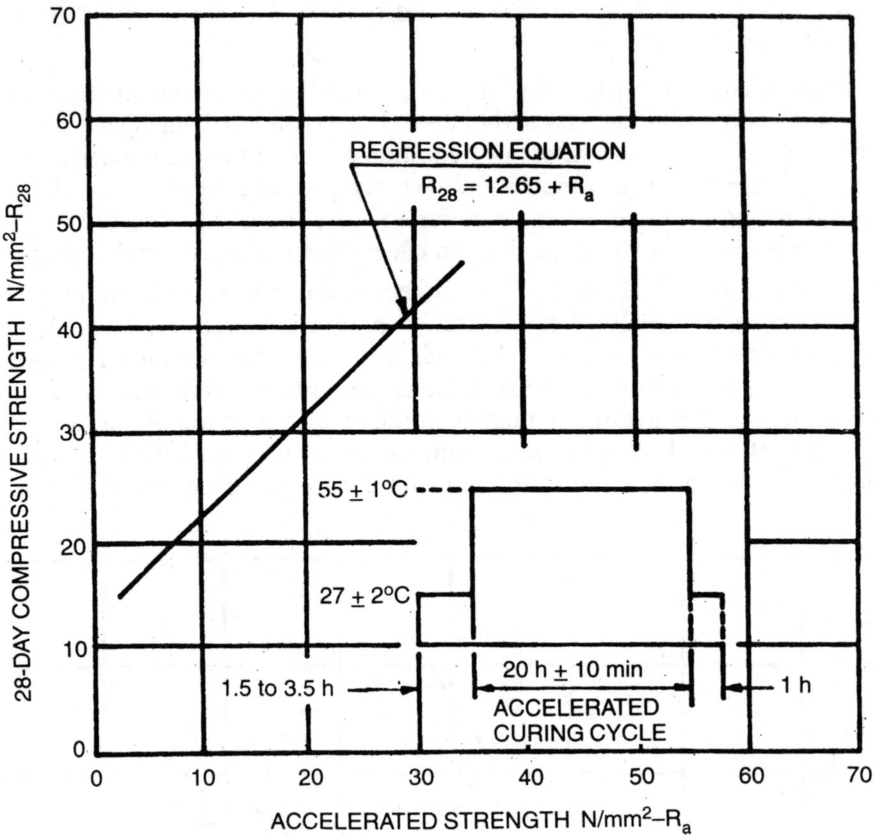

Fig. 10.7. Warm-water method. Illustration of pressure method of test for air content. F_s = weight in kg of fine aggregate in concrete sample under test; S = Volume in m^3 of concrete sample (same as volume) of measuring bowl of apparatus); B = Volume in m^3 of concrete produced per batch determined in accordance with 7 of IS:1199; F_b = Total weight in kg of fine aggregate in batch; C_s = Weight in kg of coarse aggregate in concrete sample under test; and C_b = Total weight in kg of coarse aggregate in batch.

the strength of concrete after accelerated curing (by either method) is of the order of 50 per cent of that obtained on normally cured, 28 days old specimens. When results of concrete with specific ingredients and mix proportions are considered, the dispersion of results is considerably reduced, and the coefficient of variation of results from accelerated curing methods may, in that case, be expected to be of the same order as obtained in normally cured, 28 days conventional tests.

Although the tests have shown that the correlation between results from accelerated curing method and normally cured 28 days conventional tests is not materially affected by the chemical composition, fineness and strength of cement, the mix proportions or use of some indigenous admixtures, it is preferable to establish the actual correlation under site conditions for the specific materials and mix proportions to be adopted, for use in each case.

EXPERIMENT 9

METHOD OF DETERMINING AIR CONTENT OF FRESHLY MIXED CONCRETE BY THE PRESSURE METHOD

Objective

To determine the air content of freshly mixed concrete by the pressure method.

Theory

This method of determining air content is considered adequate for all ordinary types of concrete and mortar, except for concretes or mortars made with highly porous aggregates, where the aggregate correction factor cannot be determined accurately by the technique found satisfactory for the usual types of relatively dense natural aggregates.

Apparatus

- *Measuring bowl* : A flanged cylindrical bowl, preferably of steel or hard metal not readily attacked by the cement paste, having a diameter equal to 1 to 1.25 times the height. The outer rim and upper surface of the flange, as well as the interior surfaces of the bowl, shall be smooth-machined surfaces. The minimum size of the container shall be a function of the size of a coarse aggregate in the concrete sample. Container shall be at least as large as is specified in Table 10.1, depending on the size of the aggregate used in the concrete.

  Table 10.1. Minimum size of containers corresponding to nominal maximum size of aggregate

Minimum size of container (m^3)	*Nominal maximum size of aggregate* (*mm*)
0.005	38
0.01	75
0.1	150

  The bowl shall be pressure tight and sufficiently rigid to limit the expansion factor 'D' (Calculation given in the subsequent paras) of the apparatus assembly to not more than 0.1 per cent of the air content on the standpipe indicator scale when under the normal operating pressure.

  Note : Large container may be used for large samples of concrete in order to reduce errors in sampling.

- *Conical cover assembly* : The flanged cover, preferably of steel or hard metal not readily attacked by the cement paste, shall have interior surfaces inclined not less than 30°C from the horizontal. The outer rim and lower surface of the flange and the sloping interior surface shall be pressure-tight and sufficiently rigid to limit the expansion factor of the apparatus assembly as prescribed under measuring bowl. The cover shall be fitted with a standpipe which may be a graduated precision bore glass tube or may be made of metal of uniform bore with a glass water gauge attached.

The graduations for a suitable range in air content shall be in per cent and tenths of a per cent as determined by the proper air pressure calibration test. The internal diameter of the standpipe shall be designed so that under the normal operating pressure the water column will be lowered sufficiently to measure air contents upto 0.1 per cent. It is suggested that approximately 25 mm lowering of water column should represent one per cent of air. The applied air pressure shall be indicated by a pressure gauge connected to the air chamber above the water column. The gauge shall have a range of twice the normal working pressure with suitable graduations. (A pressure of 0.5 to 2.0 kg/cm^2 has been used satisfactorily. However, each container shall have to be calibrated for a stated normal procedure). The cover shall be fitted with a suitable device for venting at the top of the air chamber, an air valve, and a petcock for bleeding off water as required. Suitable means for clamping the cover to the bowl shall be provided to make a pressure-tight seal without entrapping air at the joint between the flanges of the cover and bowl. A suitable hand pump shall be provided with the cover, either as an attachment or as an accessory.

- *Calibration cylinder* : The calibration cylinder shall consist of a cylindrical measure having an internal volume equal to approximately 3 to 6 per cent of the volume of the measuring bowl. A satisfactory measure may be machined from 1.6 mm brass tubing (No. 16 BG) (or proper diameter to provide the volume desired) to which a brass disc 6.5 mm in thickness is soldered to form the bottom.
- *A coil spring* : A coil spring or other means shall be provided for holding the calibration cylinder in place.
- *Spray tube* : A tube of appropriate diameter which may be an integral part of the cover assembly or which may be provided separately is so constructed that when water is added to the container, there will be a minimum of disturbance to the concrete.
- *A trowel* of the ordinary brick layer's type.
- *Tamping rod* : The tamping rod shall be of steel or other suitable material of 1.6 cm diameter, 61 cm long, and rounded at the tamping end.
- *Mallet* : A mallet with a rubber or rawhide head, weighing 250 g for containers smaller than 0.01 cu.m. capacity and 500 g or more for larger containers.
- *Strike-off bar* : A strike-off bar consisting of flat straight steel bar.
- *Funnel* : A funnel with spout fitting into the tube described under spray tube.
- *Measure* : A measure having a 2.5 or 5 litre capacity, as required to fill the indicator with water from the top of the concrete to the zero mark.

Calibration

(a) *Change in barometric pressure* caused by change in elevation or by changes of temperature and humidity, and rough handling under job conditions, will affect the calibration of pressure type apparatus for determination of air content. The steps described under this clause are pre-requisites for the final calibration test to determine the operating pressure P on the pressure gauge as described hereunder. Normally, this calibration need be made only once (at the time of the initial calibration), or only occasionally to check volume constant of the calibration cylinder and measuring bowl. On the other hand, the calibration test described in 'g' below must be made as frequently as necessary, to ensure that the proper gauge pressure P is being used in tests for the air content of concrete. Moreover, a change in elevation of more than 183 m from the location at which the apparatus was last calibrated will require calibration in accordance with 'g'.

(b) *Calibration of calibration cylinder* : The weight of water 'w' (in grammes) required to fill the calibration cylinder shall be accurately determined using a scale sensitive to 0.5 g.

(c) *Calibration of measuring bowl* : The weight of water 'w' (in grammes) required to fill the measuring bowl shall be determined using a scale sensitive to 0.1 per cent of the weight of the bowl filled with water. A glass plate is slid carefully over the flange of the bowl in such a manner as to ensure that the bowl is completely filled with water. A thin film of cup grease smeared on the flange of the bowl will make a water-tight joint between the glass plate and the top of the bowl.

(d) *Determination of constant R* : The constant R represents the volume of the calibration cylinder expressed as percentage of the volume of the measuring bowl. Calculate 'R' as follows :

$$R = \frac{100\,w}{w}$$

(e) *Determination of expansion factor D* : The expansion factor 'D' for any given apparatus assembly shall be determined by filling the apparatus with water only (making certain that all entrapped air has been removed and the water level is exactly on the zero mark), and applying an air pressure approximately equal to the operating pressure P, determined by the calibration test described in 'g'. The amount by which the water column is lowered shall be the equivalent expansion factor 'D' for that particular apparatus and pressure.

(f) *Determination of calibration factor 'k'* : The calibration factor 'k' is the amount by which the water column shall be depressed, during the calibration procedure to obtain the gauge pressure required to make the graduations on the glass tube correspond directly to the percentage of air introduced into the measuring bowl by the calibration cylinder where the bowl is level full of water. Calculate 'k' as follows :

$$K = 0.98\ RD$$

(g) *Calibration test to determine operating pressure, P, on pressure gauge* : If the rim of the calibration cylinder contains no recesses or projections, it shall be fitted with three or more spacers equally spaced around the circumference. Invert the cylinder and place it at the centre of the dry bottom of the measuring bowl. The spacers shall provide an opening for flow of water into the calibration cylinder when pressure is applied. Secure the inverted cylinder against displacement and carefully lower the conical cover. After the cover is clamped in place, carefully adjust the apparatus assembly to a vertical position and add water at air temperature, by means of the tube and funnel, until it rises above the zero mark on the standpipe. Close the vent and pump air into the apparatus to the approximate operating pressure. Incline the assembly about 30° from the vertical and using the bottom of the bowl as a pivot, describe several complete circles with the upper end of the standpipe, simultaneously tapping the cover and sides of the bowl lightly to remove any entrapped air adhering to the inner surfaces of the apparatus. Return the apparatus to a vertical position, gradually release the pressure (to avoid loss of air from the calibration cylinder) and open the vent. Bring the water level exactly to the zero mark by bleeding water through the petcock in the top of the conical cover. After closing the vent, apply pressure until the water level has dropped an amount equivalent to about 0.1 to 0.2 per cent of air more than the value of the calibration factor k, determined as described in 'f'. To relieve local restraints, lightly tap the sides of the bowl, and, when the water level is exactly at the value of the calibration factor k, read the pressure P, indicated by the gauge and record to the nearest 0.01 kg/cm^2. Gradually release the pressure and open the vent to determine whether the water level returns to the zero mark when the slides of the bowl are tapped lightly (failure to do so indicates loss of air from the calibration cylinder or loss of water due to a leak in the assembly). If the water level fails to return to within 0.05 per cent air of the zero mark and no leakage beyond a few drops of water is found, some air probably was lost from the calibration cylinder. In this case, repeat the calibration procedure step by step from the beginning of this paragraph. If the leakage is more than a few drops of water, tighten the leaking joint before repeating the calibration pressure. Check the indicated pressure reading promptly by bringing the water level exactly to zero mark, closing the vent, and applying the pressure 'P', just determined. Tap the gauge lightly with a finger. When the gauge indicates the exact pressure P, the water column should read the value of the calibration factor k, used in the first pressure application within about 0.05 per cent of air.

(h) *Determination of aggregate correction factor* : The aggregate correction factor shall be determined on a combined sample of fine and coarse aggregates as specified hereunder and illustrated in Fig. 10.8(a).

The weights of fine and coarse aggregates present in the volume S, of the sample of fresh concrete whose air content is to be determined, shall be determined as follows :

$$F_s = \frac{S}{B} \times F_b$$

$$C_s = \frac{S}{B} \times C_b$$

$A_1 = (h_1 \times h_2)$ when bowl contains concrete as shown in Fig. 10.8(a), when bowl contains only aggregate and water $h_1 - h_2 = G$ (aggregate correction factors), $A_1 - G = A$ (air content, percentage by volume of concrete).

Mix representative samples of fine aggregate, of weight F_s and coarse aggregate of weight C_s, and place in the measuring bowl filled one-third full of water. Add the mixed aggregate a small amount at a time, until all the aggregate, is inundated. Add each scoopful in a manner that will entrap as little air as possible and remove accumulations of foam promptly. Tap the sides of the bowl and lightly rod the upper layer of the aggregate about ten times and stir after each addition of fine aggregate of eliminate entrapped air.

When all of the aggregate has been placed in the bowl and inundated for at least 5 minutes, strike off all foam and excess water and thoroughly clean the flanges of both bowl and conical cover so that when the cover is clamped in place, pressure-tight seal is obtained. Complete the test as described in procedure for determining air content of concrete. The aggregate correction factor G is equal to $h_1 - h_2$ as determined in the tests on the aggregate.

Note : The aggregate correction factor will vary with different aggregates. It can be determined only by test, since apparently it is not directly related to absorption of the particles. The test can be easily made and shall not be ignored. Ordinarily the factor remains reasonably constant for given aggregates, but an occasional check test is recommended.

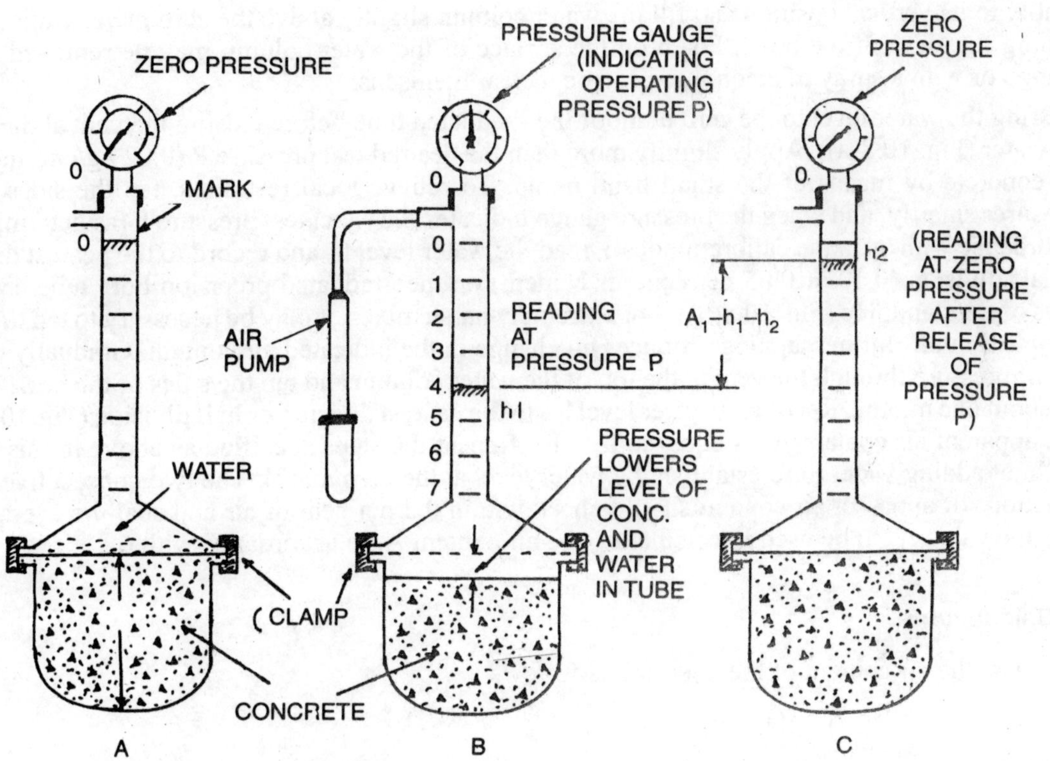

Fig. 10.8. Illustration of pressure method of test for air content.

Procedure for determining air content of concrete

Place a representative sample of the concrete in the measuring bowl in three equal layers, consolidating each layer by rodding the bowl. Vibration may be substituted for rodding and by tapping the sample when the air content of concrete placed by vibration is to be determined. When the concrete is to be placed by rodding, consolidate each layer of concrete by about 25 strokes of the tamping rod evenly distributed over the cross-section. Follow the rodding of each layer by tapping the sides of the bowl smartly 10 to 15 times with the mallet until the cavities left by rodding are levelled out and no large bubbles of the air appear on the surface of the rodded layer. In rodding the first layer, the rod shall not forcibly strike the bottom of the bowl. In rodding the second and final layers, only enough force shall be used to cause the rod to penetrate the surface of the previous layer. Slightly overfill the bowl with the third layer and, after rodding or vibration, remove the excess concrete by sliding the strike-off bar across the top flange with a sawing motion until the bowl is just level full.

Thoroughly clean the flanges of the bowl and of the conical cover so that when the cover is clamped in place, a pressure-tight seal will be obtained. Assemble the apparatus and add water over the concrete by means of the tube until it rises to about half way mark in the standpipe. Incline the apparatus assembly about 30° from vertical end, using the bottom of the bowl as a pivot, describe several complete circles with the upper end of the column simultaneously tapping the conical cover lightly to remove any entrapped air bubbles above the concrete sample. Return the apparatus assembly to its vertical position and fill the water column slightly above the zero mark, while lightly tapping the sides of the bowl. Foam on the surface of the water column may be removed with a syringe or with a spray of alcohol to provide a clear meniscus.

Bring the water level to the zero mark of the graduated tube before closing the vent at the top of the water (Fig. 10.8(a)). Apply slightly more than the desired test pressure P (0.02 kg/cm^2 more) to the concrete by means of the small hand pump. To relieve local restraints, tap the sides of the measures smartly, and when the pressure gauge indicates the exact test pressure P (as determined in accordance with 'g' in the calibration test), read the water level h_1 and record to the nearest division or half division (0.10 or 0.05 per cent air content) on the graduated precision bore tube or gauge glass of the standpipe (Fig. 10.8(b)). For extremely harsh mixes, it may be necessary to tap the bowl vigorously until further tapping produces no change in the indicated air content. Gradually release the air pressure through the vent at the top of the water column and tap the sides of the bowl lightly for about one minute. Record the water level h_2, to the nearest division or half division (Fig. 10.8(c)). The apparent air contant A_1 is equal to $h_1 - h_2$. Repeat the steps specified as above in this clause (without adding water to re-establish the water level at the zero mark). The two consecutive determinations of apparent air content should check within 0.2 per cent of air and shall be averaged to give the value A_1 to be used in calculating the air content A, in accordance with 6.

6. Calculations

Calculate the air content of the concrete as follows :

$$A = A_1 - G$$

where :

A = air content, percentage by volume of concrete.

A_1 = apparent air content, percentage by volume of concrete, and

G = aggregate correction factor, percentage by volume of concrete.

EXPERIMENT 10

METHOD OF DETERMINING CEMENT CONTENT OF HARDENED PORTLAND CEMENT CONCRETE

Objective

The method of test specifies the procedure for determining the cement content of hardened portland cement concrete except those containing certain aggregates or admixtures which liberate soluble silica under the conditions of the test, such as slags, diatomites and sodium silicate.

Reagents

- *Hydrochloric acid* - approximately 3.3 N. Add 200 ml of hydrochloric acid (specific gravity 1.19) to 600 ml of distilled water.

- *Sodium hydroxide* - approximately 1 N. Dissolve 20 g of hydroxide in 200 ml of water and dilute to a volume of 500 ml
- *Hydrofluoric acid* - 40 per cent.
- *Sulphuric acid* - specific gravity 1.84.

Preparation of sample

Every precaution shall be taken to have the sample of concrete used for analysis truly representative of the material under consideration. Several portions weighing at least 5 kg each shall be taken to avoid all but slight inequalities of the concrete mix. These portions shall then be broken up, crushed in a suitable machine to about one centimetre size and reduced to a fineness of approximately IS Sieve 10 to IS sieve 8 in a ball mill, disc pulveriser or by any other suitable device. Care shall be taken that the finer fractions of the broken sample, which are richer in cement, are not discarded or lost. After thorough mixing and quartering, a portion approximately 100 g shall be taken and carefully freed, by means of a strong magnet, from particles of metallic iron abraded from the pulveriser ball mill. The clean sample shall then be dried at 105°C for at least 2 hours.

Procedure

Weigh into each of three 250 ml beakers, not less than a 2 g portion of the prepared sample. Moisten with a stream of hot water, while stirring to prevent adhesion to the beaker or the formation of lumps in the mass. Slowly add 100 ml of 3.3 N hydrochloric acid and stir thoroughly. The lumps which tend to form should be reduced with the glass rod. After the evolution of carbon dioxide has ceased and the reaction is apparently complete, heat gently for a few minutes and allow the contents of the beaker to settle. Decant through an ignited and weighed Gooch crucible which contains a mat of short asbestos shreds, practically insoluble in hydrochloric acid and thick enough to be opaque to light. Once the filtration has begun, care shall be taken so that the mat and accumulated residue do not dry out completely until the filtration process is complete. Regulate the suction so as to maintain a rapid rate of dropping during the greater part of the filtration. Retain as much of the residue in the beaker as possible. Wash by decantation twice with hot water. Add 75 ml of 1 N sodium hydroxide to the residue while stirring and heat to about 75°C. Decant as before and wash twice with hot water. Transfer the residue to the crucible and wash with at least 60 ml of hot water.

The filtrate now contains the silica in the form of silicic acid in true solution or in suspension in the hydrochloric acid medium. If the aggregates of the original sample are largely calcareous or dolomitic, add 10 ml of hydrochloric acid (specific gravity 1.19) to the solution. Transfer to a suitable beaker with several rinsings of the filter-flask. Evaporate to dryness with great care to minimize spattering, bake at not over 120°C for one hour, moisten with hydrochloric acid (specific gravity 1.19), evaporate and bake again and take for filtration in 75 ml of 2 N or 3 N hydrochloric acid heated to boiling. Filter through an ashless filter paper and wash the residue with 50 ml of hot 1 N hydrochloric acid and then with hot water until the washings are free from chloride. Repeat the evaporation and filtering processes to recover the small amounts of silica dissolved and add these to the first residue. Determine the silica present in the sample by treatment with hydrofluoric and sulphuric acids in accordance with the procedure given in Appendix A of IS : 269.

Correction factor

When the aggregates, used in the concrete being analyzed, are available, a blank test shall be run on these aggregates to determine their content of silica, soluble under the conditions of the test. This content of the soluble silica shall then be used as a correction factor and be subtracted from the total soluble silica found in the concrete, the difference being due to the cement contained in the specimen.

Calculation

The percentage of cement in the sample shall be calculated by dividing the percentage of silica found by the factor 0.2140, provided the silica content of the cement is not known to be different from this value. When possible, the known value shall be taken as the factor.

CHAPTER 11

Non-Destructive Tests and Latest Techniques

METHOD 1—ULTRASONIC PULSE TECHNIQUE

Objective

The ultrasonic pulse velocity method could be used to establish :

(a) the homogeneity of the concrete,
(b) the presence of cracks, voids and other imperfections,
(c) changes in the structure of the concrete which may occur with time,
(d) the quality of the concrete in relation to standard requirements,
(e) the quality of one element of concrete in relation to another, and
(f) the values of dynamic elastic modulus of the concrete.

Principle

The ultrasonic pulse is generated by an electro-accoustical transducer. When the pulse is induced into the concrete from a transducer, it undergoes multiple reflections at the boundaries of the different material phases within the concrete. A complex system of stress waves is developed which includes longitudinal, shear and surface waves. The receiving transducer detects the onset of the longitudinal waves, which is the fastest.

Since the velocity of the pulses is almost independent of the geometry of the material through which they pass and depends only on its elastic properties, pulse velocity method is a convenient technique for investigating structural concrete. The underlying principle of assessing the quality of concrete that comparatively higher velocities are obtained when the quality of concrete in terms of density, homogeneity and uniformity is good. In case of poorer quality, lower velocities are obtained. If there is a crack, void or flaw inside the concrete which comes in the way of transmission of the pulses, the pulse strength is attenuated and it passes around the discontinuity, thereby making the path length longer, and hence the lower velocities are obtained. The actual pulse velocity obtained depends primarily upon the materials and mix proportions of concrete. Density and modulus of elasticity of aggregate also significantly affect the pulse velocity.

Apparatus

- Electrical pulse generator
- Transducer - one pair
- Amplifier
- Electronic timing device

Procedure

In this method, the ultrasonic pulse is produced by the transducer which is held in contact with one surface of the concrete member under test. After traversing a known path length (L) in the concrete, the pulse of vibrations is converted into an electrical signal by the second transducer held in contact with the other surface of the concrete member on an electronic timing circuit which enables the transit time (T) of the pulse to be measured. The pulse velocity (V) is given by :

$$V = L/T$$

Once the ultrasonic pulse impinges on the surface of the material, the maximum energy is propagated at right angles to the face of the transmitting transducer and the best results are, therefore obtained when the receiving transducer is placed on the opposite face of the concrete member (direct transmission or cross probing). In many situations, two opposite faces of the structural member may not be accessible for measurements. In such cases, the receiving transducer is also placed on the same face of the concrete members (surface probing). Surface probing is not so efficient as cross probing, because the signal produced at the receiving transducer has an amplitude of only 2 to 3 per cent of that produced by cross probing and the test results are greatly influenced by the surface layers of concrete which may have different properties from that of concrete inside the structural member. The indirect velocity is lower than the direct velocity and the difference may vary from 5 to 20 per cent depending on the quality of concrete under test. For good results there should be adequate accoustical coupling between the concrete and the face of the transducer. Couplants like petroleum jelly, grease, liquid soap etc. may be used.

In view of the inherent variability in the test results, sufficient number of readings are taken by dividing the entire structure in suitable grid markings of 30 cm × 30 cm or even smaller. Each junction point of the grid becomes a point of observation. Surface probing in general gives lower pulse velocity than in case of cross probing and depending on number of parameters, the difference could be of the order of about 1 km/sec.

Observation table

S.No.	*Method of testing direct/indirect/ semi-direct*	*Distance between probe points (cm)*	*Time of pulse (sec)*	*Pulse velocity (cm/sec) or (km/sec)*
1				
2				
3				

Average pulse velicity in km/sec = ...

Remarks : ....

Interpretation of results

The guidelines to be followed to assess the quality of concrete are given as under.

Table 11.1. Velocity criterion for concrete quality grading

S. No.	*Pulse velocity by cross probing/ direct method (km/sec)*	*Concrete quality grading*
1	Above 4.5	Excellent
2	3.5 to 4.5	Good
3	3.0 to 3.5	Medium
4	Below 3.0	Doubtful

Note : In case of "doubtful" quality it may be necessary to carry out further tests.

The dynamic Young's modulus of elasticity (E) of the concrete can be determined from the pulse velocity and the dynamic Poisson's ratio (μ) using the following relationship :

$$E = \frac{\rho\,(1+\mu)\,(1-2\mu)}{1-\mu}\,v^2$$

where :

E = dynamic Young's modulus of elasticity in MPa.
ρ = density in kg/m^3, and
v = pulse velocity in m/sec

The above relationship may be expressed as :

$$E = \rho f\,[\mu]\,v^2$$

where

$$f(\mu) = \frac{(1+\mu)\,(1-2\mu)}{1-\mu}$$

The value of the dynamic Poisson's ratio varies from 0.20 to 0.35, with 0.24 as average. The dynamic Poisson's ratio may be obtained from measurements on concrete test beams of the pulse velocity (v) alongwith length (l) of the beam and the fundamental resonant frequency (n) of the beam in longitudinal mode of vibration. From these measurements, the factor (μ) is calculated by the relation :

$$f(\mu) = \frac{(2\,nl)^2}{v^2}$$

where

n = fundamental resonant frequency in cycles per second, and
l = length of specimen in 'm'

Precautions

(a) There should be adequate accoustical coupling between the concrete surface and the transducer.
(b) If the surface of concrete is rough, it is essential to smoothen the surface.
(c) High frequency transducers should be used for short path lengths and low frequency transducers for longer path lengths.
(d) For concretes having maximum nominal size of aggregate as 20 mm or less minimum path length should be 100 mm and for concrete having maximum nominal size of aggregate between 20 mm to 40 mm the minimum path length should be 150 mm.

Influence of test conditions

(i) Influence of surface conditions and moisture content of concrete

Surface of the concrete should be smooth, to ensure good accoustical coupling between the transducer against the concrete surface. If the surface is rough and uneven, it is necessary to smoothen the surface. In general, the pulse velocity through concrete increases with increased moisture content of concrete. This influence is more for low strength concrete than high strength concrete. The pulse velocity of saturated concrete may be upto 2 per cent higher than that of similar dry concrete.

(ii) Influence of path length, shape and size of the concrete member

As concrete in inherently heterogeneous, it is essential that path lengths be sufficiently long so as to avoid any error introduced due to its heterogeneity. In field work, this does not pose any difficulty as the pulse velocity measurements are carried out on thick structural concrete members. However, in the laboratory where generally small specimens are used, the path length can affect the pulse velocity readings.

The shape and size of the concrete member do not influence the pulse velocity unless the least lateral dimension is less than a certain minimum value, for example the minimum lateral dimension of about 80 mm for 50 KHz natural frequency of the transducer. Table 11.2 gives the guidance on the choice of the transducer natural frequency for different path lengths and minimum transverse dimension of the concrete members.

Table 11.2. Natural frequency of transducers for different path lengths

Path length (mm)	*Natural frequency of transducer (KHz)*	*Minimum transverse dimensions of members (mm)*
Upto 500	150	25
500-700	≥ 60	70
700-1500	≥ 40	150
Above 1500	≥ 20	300

(iii) Influence of temperature of concrete

Variations of the concrete temperature between 5 and 30°C do not significantly affect the pulse velocity measurements of concrete. At temperatures between 30 to 60°C there can be reduction in pulse velocity upto 5 per cent. Below freezing point temperature, the free water freezes within concrete, resulting in an increase in pulse velocity upto 7.5 per cent.

(iv) Influence of stress

When concrete is subjected to a stress which is abnormally high for the quality of the concrete, the pulse velocity may be reduced due to the development of the micro-cracks. This influence is likely to be greatest when the pulse path is normal to the predominant direction of the planes of such micro-cracks. This occurs when the pulse path is perpendicular to the direction of a uniaxial compressive stress in a member. This influence is generally insignificant unless the stress is greater than about 60 per cent of the ultimate strength of concrete.

(v) Effect of reinforcing bars

The pulse velocity measured in reinforced concrete in the vicinity of reinforcing bars is usually higher than in plain concrete of the same composition. This is because, the pulse velocity in steel is 1.2 to 1.9 times the velocity in plain concrete and, under certain conditions, the first pulse to arrive at the receiving transducer travels partly in concrete and partly in steel. The apparent increase in pulse velocity depends upon the proximity of the measurements to the reinforcing bar, the diameter and number of the bars and their orientation with respect to the path of propagation.

METHOD 2—SCHMIDT REBOUND HAMMER TEST

Objective

The rebound hammer method could be used for :

(i) assessing the likely compressive strength of concrete with the help of suitable co-relations between rebound index and compressive strength,

(ii) assessing the uniformity of concrete,

(iii) assessing the quality of the concrete in relation to standard requirements, and

(iv) assessing the quality of one element of concrete in relation to another.

Principle

When the plunger of rebound hammer is pressed against the surface of the concrete, the spring controlled mass rebounds and the extent of such rebound depends upon the surface hardness of concrete. The surface hardness and therefore the rebound is taken to be related to the compressive strength of the concrete. The rebound is read off along a graduated scale and is designated as the rebound number or rebound index.

Apparatus

- Rebound hammer (Fig. 11.1).

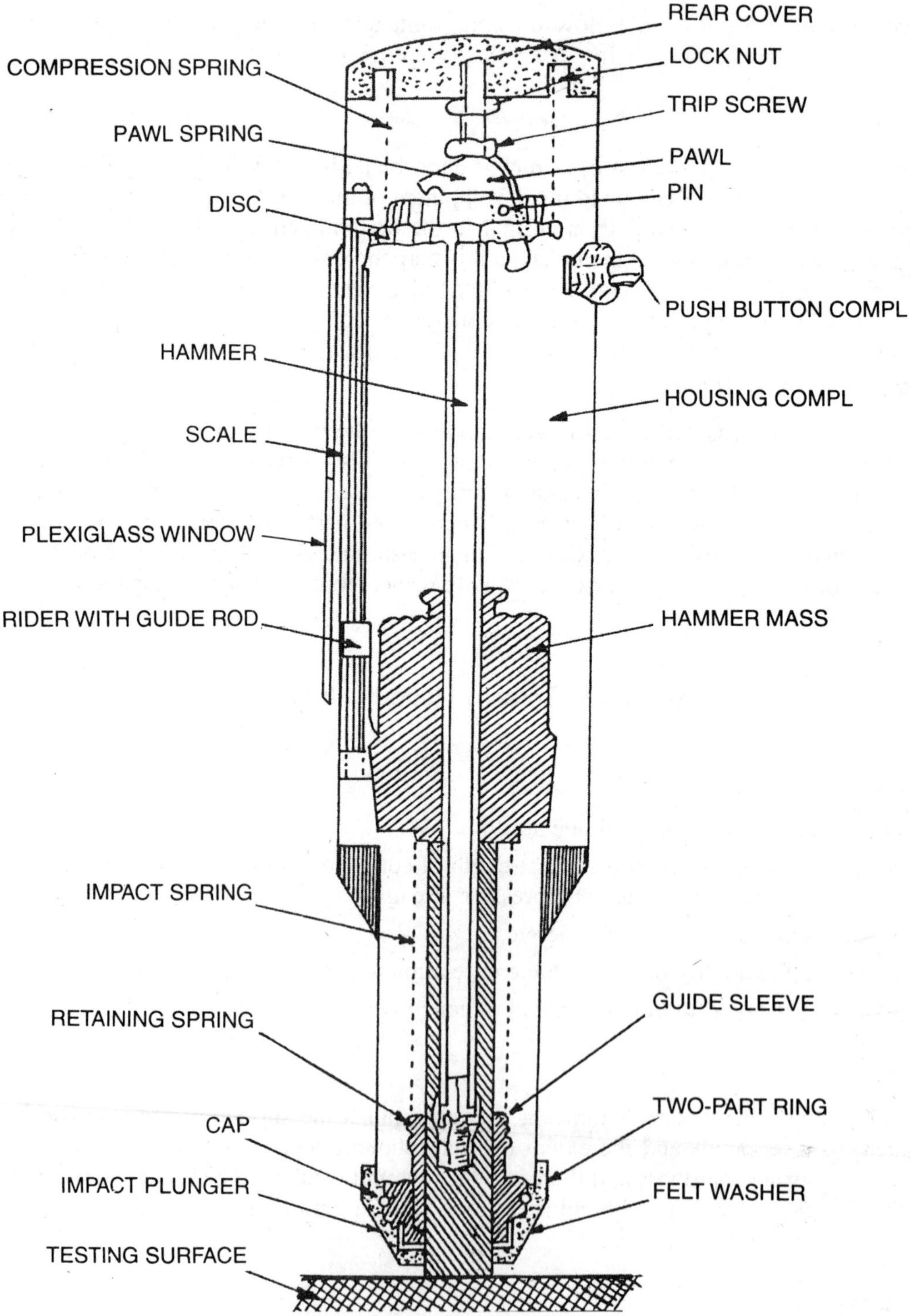

Fig. 11.1. Rebound hammer.

Table 11.3. Impact energy for rebound hammers for different applications

S. No.	*Application*	*Approximate impact energy required for the rebound hammers (Nm)*
(i)	For testing normal weight concrete	2.25
(ii)	For light-weight concrete or small and impact sensitive parts of concrete	0.75
(iii)	For testing mass concrete, for example in roads, air-field pavements and hydraulic structures	30.00

Procedure of obtaining correlation between compressive strength of concrete and rebound number

The most satisfactory way of establishing a correlation between compressive strength of concrete and its rebound number is to measure both the properties simultaneously on concrete cubes. The concrete cube specimens are held in a compressive testing machine under a fixed load, measurements of rebound number taken and then the compressive strength determined as usual. The fixed load required is of the order of 7 N/mm^2 when the impact energy of the hammer is about 2.2 Nm. The load should be increased for calibrating rebound hammers of greater impact energy and decreasing for calibrating rebound hammers of lesser impact energy.

The test specimens should be as large a mass as possible in order to minimise the size effect on the test results of a full scale structure. 150 mm cube specimens are preferred for calibrating rebound hammers of lower impact energy (2.2 Nm), whereas for rebound hammers of higher impact energy, for example, 30 Nm, the test cubes should not be smaller than 300 mm. If the specimens are wet cured, they should be removed from wet storage, and kept in the laboratory atmosphere for about 24 hours before testing. To obtain a correlation between rebound numbers and strength of wet cured and wet tested cubes, it is necessary to establish a correlation between the strength of wet tested cubes and the strength of dry tested cubes on which rebound readings are taken. A direct correlation between rebound numbers on wet cubes and strength of wet cubes is not recommended.

Only the vertical faces of the cubes as cast should be tested. At least nine readings should be taken on each of the two vertical faces accessible in the compression testing machine when using the rebound hammers. The points of impact on the specimen must not be nearer an edge than 20 mm and be not less than 20 mm from each other. The same points must not be impacted more than once.

Procedure

(i) Smoothen the concrete surface with the help of the carborundum stone removing any loosely adhered scale to the concrete surface.

(ii) Rough surfaces resulting from incomplete compaction, loss of grout, spalled or tooled surfaces do not give reliable results and should be avoided.

(iii) The point of impact should be at least 20 mm away from any edge or shape discontinuity.

(iv) The rebound hammer measurements can be taken at any angle, vertically upwards, downwards, horizontal, or any intermediate angle as the situation demands.

(v) Six readings are taken about each point of observation and the average of these readings after deleting outliners becomes the rebound index for the point of observation.

Observation table

S. No.	*Rebound number (6 readings)*	*Location of observation*	*Compressive strength (kg/cm^2)*
1			
2			
3			

Remarks : ....

Interpretation of results

It has been found out that the rebound indices are indicative of compressive strength of concrete to a limited depth from the surface. If the concrete in a particular number has internal micro-cracking, flaws or heterogeneity across the cross-section, rebound hammer indices will not indicate the same. Accuracy of prediction of compressive strength of concrete by rebound hammer is $\pm$ 25 per cent.

Influence of test conditions

1. Influence of type of cement

Concretes made with high alumina cement can give strengths 100 per cent higher than that with ordinary portland cement. Concretes made with supersulphated cement can give 50 per cent lower strength than that with Ordinary Portland Cement.

2. Influence of type of aggregate

Different types of aggregate used in concrete give different correlations between compressive strength and rebound numbers. Normal aggregates such as gravels and crushed rock aggregates give similar correlations, but concrete made with light-weight aggregates require special calibration.

3. Influence of surface condition and moisture content of concrete

The rebound hammer method is suitable only for close texture concrete. Open texture concrete typical of masonry blocks, honeycombed concrete or no fines concrete are unsuitable for this test. All correlations assume full compaction, as the strength of partially compacted concrete bears no unique relationship to the rebound numbers. Trowelled and floated surfaces are harder than moulded surfaces, and tend to overestimate the strength of concrete.

A wet surface will give rise to underestimation of the strength of concrete calibrated under dry conditions. In structural concrete, this can be about 20 per cent lower than in an equivalent dry concrete.

4. Influence of curing and age of concrete

The relationship between hardness and strength varies as a function of time. Variations in initial rate of hardening, subsequent curing and conditions of exposure also influence the relationship. Separate calibration curves are required for different curing regimes but the effect of age can generally be ignored for concrete between 3 days and 3 months old.

5. Influence of carbonation of concrete surface

Carbonated concrete gives an overestimate of strength which in extreme cases can be up to 50 per cent. It is possible to establish correction factors by removing the carbonated layer and testing the concrete with the rebound hammer on the uncarbonated concrete.

METHOD 3—INFORMATION ON SOME LATEST NON-DESTRUCTIVE EQUIPMENTS/TESTS

Corrosion analysing instrument

This instrument is used for the non-destructive detection of corrosion in the reinforcement bars of concrete building cements. It discovers rust before it becomes visible and has caused disastrous damage.

Resonant frequency tester

This test method covers measurement of the fundamental transverse, longitudinal, and torsional frequencies of concrete for the purpose of calculating dynamic Young's modulus of rigidity (modulus of elasticity in shear) and dynamic Poisson's ratio.

Pull-out tester

In situ pull-out testing of concrete to assess strength is based on the concept that the strength of concrete is related to the maximum tensile load applied to an insert before the concrete fails. Both the Lok-test and Capo-test are now widely accepted as standard methods to determine early formwork stripping times and in-situ strength of concrete.

Crack measurement

Crack width measurement may be used when assessing concrete members, masonry walls and other types of structure. Cracks can appear in a structure for a variety of reasons including frost action, sulphate expansion, natural shrinkage, expansion or movement and corrosion of reinforcement.

Windsor probe (for compressive strength of fresh and mature concrete)

- Fast and economical to use.
- Determines the developing strength of maturing concrete, improves safety, ensures quality and reduces costs.
- Monitors the strength of mature concrete as it ages to determine the need for maintenance and restoration.
- Allows for early determination of concrete strength to :
  (i) strip forms safely,
  (ii) start post tensioning,
  (iii) maintain or improve construction schedule,
  (iv) assure the quality of the mix.

Rapid test method for concrete content of mortar

The test takes only two minutes and is primarily for use with fresh mortar but can be used for set uncarbonated mortars. An accurate weight of mortar is stirred into a standard amount of concentrated acid. The acid reacts with mortar and resultant temperature rise is proportional to the amount of cement in the mix. The test kit gives a temperature rise of 28°C with a 1 : 4 (by volume) cement: sand ratio.

Flowmeter

The flowmeter provides a fast means of establishing the flow, consistency and workability of fresh concrete with or without superplasticizers. It also measures the temperature of fresh concrete.

Poroscope

The non-destructive test is carried out in the field to determine the air and water permeability of concrete using the Figg Techniques. Each test can be completed in only a few minutes and gives reliable reproducible results. The test enables meaningful concrete durability predictions to be made.

Pile integrity tester

A quick and inexpensive method to check the continuity of installed foundation piles. The method will detect pile defects including cracks, soil inclusions, variation in pile diameter and pile length. It may be used for cast-in-situ piles and for precast piles.

M-meter

This instrument is micro-processor based and it applies the relationship of time and temperature to concrete strength for monitoring poured concrete at the site. This improves safety, ensures quality and reduces cost.

PPRmeter

This instrument measures the strength of new or existing construction materials in situ utilising the established principle of resistance to penetration.

Table 11.4. List of equipment for concrete laboratory

S.No.	*Item*	*Quantity*	*Rate*
1.	Blaine-air permeability apparatus for testing of fineness of cement as specified by IS-4031	One	2000
2.	Brass split-cylindrical mould 30 mm dia. and 30 mm high for soundness test by 'Le Chatelier Method' as specified by IS : 4031	One	500
3.	Vicat's apparatus with plunger 10 mm dia., 1 mm square needle, and 1 mm square needle with 5 mm dia., attachment for conducting test for consistency of standard cement paste, initial and final setting time test as specified by IS : 4031	One	1000
4.	Vibration machine along with poking rod, for conducting compressive strength of cement as per specifications contained in IS : 4031	One	10500
5.	Standard moulds for cement mortar cube specimens of 50 cm^2 face area for compressive strength of cement as per specifications contained in IS : 4031	One	250
6.	Compression testing machine 100 ton capacity, electrical cum-hand operated working on 3-phase 440 V	One	50000

(Contd.)

S.No.	Item	Quantity	Rate
7.	Set of sieves for the following sizes consisting of brass frames and also conforming to the IS : 460 :		
	(a) Square-hole, perforated plate of 30 cm dia. frame with sieve designations. 80 mm, 63 mm, 50 mm, 40 mm, 20 mm, 16 mm, 12.5 mm, 10 mm, 6.3 mm, 4.75 mm	One each	340
	(b) Fine-mesh wire-cloth sieves of 20 cm dia. frame with sieve designations. 2.36 mm, 1.18 mm, 600 micron, 300 micron, 150 micron, 75 micron	One each	240
	(c) Square hole perforated plate of 20 cm dia. frame with sieve designations 10 mm, 4.75 mm	Each	140
8.	Metal gauge for measuring flakiness index as specified in IS : 2386 - Part I	One	800
9.	Length gauge for measuring elongation index as specified in IS : 2386 - Part I	One	750
10.	Electrically heated oven of chamber size 60 cm × 60 cm with forced air-circulation with thermostat and automatic temperature control. Temperature range 0-150°C	One	10000
11.	Cylindrical metal measure of 3, 15 and 30 litre capacity, along with a tamping rod, 60 mm dia. and 60 cm long, rounded at one end and other specifications as per IS : 2386 - Part III for measuring bulk-density of aggregate	1 set of 3	3000
12.	Impact testing machine of the specifications contained in IS : 2386 - Part IV along with cylindrical metal measure of dia. 75 mm and depth 50 mm and tamping rod 10 mm dia., 230 mm long and rounded at one end. Impact testing machine shall have automatic arrangement for recording number of blows	One	7000
13.	Mould for conducting slump test with 20 cm bottom dia., 10 cm top dia. and height 30 cm with other specifications as contained in IS : 1199, along with tamping rod 16 mm dia., 60 cm long and rounded at one end	One	650
14.	Compacting factor test apparatus of essential dimensions as specified IS : 1199	One	8000
15.	Cylindrical metal measure of the nominal capacity 0.01 m^3 with inside dia. 250 mm, inside, height 280 mm with the minimum thickness of metal 4 mm with other specifications as contained in IS : 1199 for determining unit weight of concrete	One	950
16.	Cube moulds of size 15 cm × 15 cm × 15 cm as specified in IS : 516	One	300
17.	Cylindrical moulds with 15 cm internal dia., and height 30 cm and with other specifications as per IS : 516	One	250
18.	Beam mould of the size 15 × 15 × 70 cm and with the other specifications as contained in IS : 516	One	1500
19.	Concrete mixer 100 litre capacity tilting type mounted on wheels, fitted with diesel engine	One	15000
20.	Immersion type needle vibrator 40 mm dia. with other specifications as contained in IS : 2505	One	6500
21.	25 mm size riffler made of G.I. sheet with receivers, one scoop and iron stand	One	1500
22.	Spring balance of capacity 200 kg reading upto 1 kg	One	1000
23.	Platform weighing machine of capacity 300 kg with min. reading of 0.5 kg. Avery make or equivalent	One	12000
24.	Wire-basket of not more than 63 mm mesh as per IS : 2386 (Part III)	One	400
25.	Table pan balance 10 kg capacity sensitive upto 1 gm	One	1500
26.	Table pan balance 5 kg capacity sensitive upto 0.5 gm	One	1400
27.	Physical balance	One	800
28.	Weight boxes with range of weights from mg weight to 500 gm	One	700
29.	Weight boxes with range of weights from mg weights to 200 gm	One	650

Note : Rate/price of equipment mentioned in the manual is an approximate amount. Exact amount can be obtained from the manufacturers.

Table 11.5. Consumable materials required for concrete laboratory

S.No.	*Item*	*Quantity*	*Rate*
1.	Trowels : Large	6	50
	Small	6	25
2.	Enamel trays 69 cm × 30 cm	6	150
3.	Spades	4	75
4.	Shovels	4	100
5.	Glass measuring cylinders graduated : 1000 cc	6	150
	500 cc	6	100
6.	Beakers 250 cc	6	95
7.	Glass jars 25 cm × 10 cm with well ground flat edges for surface moisture test	2	100
8.	Thermometer range 0-100°C	2	100
	0-55°C	2	150
9.	G.I. buckets 10 litres	4	84
	15 litres	4	90
10.	Funnels	2	100
11.	G.I. sheet 2.4 × 10 m	1	600
12.	Scoops - Big size	4	100
	Small size	4	90
13.	Hammer 3.0 lbs	3	100
14.	Electrical heater thermostatically controlled	1	750
15.	Aluminium pot with cover sufficiently large to hold Le-Chatelier mould for soundness test	2	200
16.	Plastic mugs one litre capacity	2	20
	two litre capacity	2	30
	three litre capacity	2	55
17.	Standard sand as per IS : 650	100 kg	5000 / ton
18.	Open cylindrical tin boxes to store about 25 litres of aggregate	6	155
19.	Hammers 3 lbs	3	100
20.	Mobil oil - low-grade	5 litres	18 / litre
21.	Petrol	8 litres	28 / litre
22.	Diesel	8 litres	12 / litre
23.	Empty tins for collecting the above material	5 Nos.	50
24.	Filter paper	1 packet	25
25.	Coarse aggregate 20 mm		
	40 mm	LS	1000 /cmm
26.	Fine aggregate	LS	500 /cmm
27.	Cement bags	12	200
28.	Hand brush 60 cm	2	100
29.	Hand brush 22.5 cm	2	70
30.	Steel brush	2	150

Bibliography

1. IS : 269, Ordinary and low heat Portland Cement.
2. IS : 383, Aggregates, coarse and fine from natural surfaces for concrete.
3. IS : 1489, Portland Pozzolana cement.
4. IS : 6909, Super sulphated cement.
5. IS : 8112, 43 grade Ordinary Portland cement.
6. IS : 12269, 53 grade Ordinary Portland cement.
7. IS : 456, Code of practice for plain and reinforced concrete.
8. IS : 1199, Sampling and analysis of concrete.
9. IS : 516, Strength of concrete.
10. IS : 2386, Methods of test for aggregates for concrete.
11. IS : 4031, Hydraulic cement, physical tests for method of chemical analysis of hydraulic cement.
12. IS : 2430, Methods of sampling of aggregates for concrete.
13. IS : 3535, Methods of sampling of hydraulic cements.
14. IRC : 44, Design concrete mixes for road pavements.
15. IRC : 44, Tentative guidelines for cement concrete mix design for road. Pavement for non-air entrained and continuously graded concrete.
16. IS : 9013, Method of making, curing and determining compressive strength of accelerated-cured concrete test specimens.
17. IS : 13311 (Part-I), Non-destructive testing of concrete - Methods of test : Ultrasound Pulse Velocity.
18. IS : 13311 (Part-II), Non-destructive testing of concrete - Methods of test : Rebound Hammer.
19. IS : 5516, Blaine's air permeability apparatus.
20. Portland Cement Association, design and control of concrete mixture, July, 1958.

21. Elvery, concrete, Practice, Vols. I and II, Asia Publishing House (1963).
22. Neville, Properties of concrete, Pitman (1968).
23. Portland Cement Association, Concrete Technology (student manual), Taraporewolla (1969).
24. Concrete Association of India, Concrete Handbook.
25. Bundes Verband der dutschen Zemintindustrie : Zement Markblatter, 1965.
26. Each Verband Zemant, Arbeitsblatter zur prifung, der Zement Zuschlagstoffe and Beton, Beton verleg Gmbh.
27. Institution of Structural Engineers, Report on Concrete Practice, Part-I, Materials and Workmanship (1963).
28. Hackley, Concrete Practice, Chato and Hindus, London (1959).
29. Road Research Lab, London, Road Note No. 4, Design of Concrete Mixes (1969).
30. Orchard Concrete Technology, Vols. I and II, Contractor Record Limited, London.
31. McIntosh, Design of Concrete Mix, Cement and Concrete Association, London.

NOTES

NOTES